Collins
complete
plumbing AND
heating

Albert Jackson and David Day

Collins

Collins Complete Plumbing and Heating
was originally created for HarperCollins Publishers by
Inklink/Jackson Day Jennings. Most of the material in
this book also appears in *Collins Complete DIY Manual*.

This paperback edition published in 2010

First published in 2008 by
Collins, an imprint of
HarperCollinsPublishers
77-85 Fulham Palace Road
Hammersmith
London W6 8JB

Collins is a registered trademark
of HarperCollins Publishers Ltd

15   14   13   12   11   10
6   5   4   3   2   1

A catalogue record for this book is available from
The British Library

ISBN 978 000 737949 1

Colour reproduction by Colourscan, Singapore
Printed and bound by Printing Express, Hong Kong

PLEASE NOTE
**Great care has been taken to ensure that the
information contained in Collins Complete Plumbing
and Heating is accurate. However, the law concerning
Building Regulations, planning, local bylaws and
related matters is neither static nor simple. A book
of this nature cannot replace specialist advice in
appropriate cases and therefore no responsibility
can be accepted by the publishers or by the authors
for any loss or damage caused by reliance upon the
accuracy of such information.**

**If you live outside Britain, your local conditions
may mean that some of this information is not
appropriate. If in doubt, always consult a qualified
electrician, plumber or surveyor.**

Authors
Albert Jackson and David Day

Contributors
David Bridle

Photographers
Airedale/David Murphy
Colin Bowling
Focus Publishing
Paul Chave
Ben Jennings
Neil Waving

Set Building/Projects for Photography
Airedale
David Bridle
Bill Brooker
Focus Publishing

Consultant
Roger Bisby

Design
Keith Miller

New Illustrations
Graham White

Illustrations
Robin Harris
John Pinder

Editors
Peter Leek
Barbara Dixon

Proofreader and Indexer
Mary Morton

Acknowledgements
The authors and publishers would like to thank the following
companies and individuals who supplied images or tools for
photography:

12 Opella Ltd; 13 Screwfix; 22 Cuprofit/IBP Group; 30 Ideal
Standard; 32 Ideal Standard; 34 Ideal Standard; 38 Saniflo Ltd;
39 Ideal Standard, Aqualisa Products Ltd; 40 Ideal Standard;
41 Mira Showers; 42 Ideal Standard, Hans Grohe Ltd; 46 Franke
UK Ltd, Rangemaster; 47 Rangemaster; 52 Albion Water
Heaters Ltd; 53 Rayotec Ltd; 54 Solartwin; 58 Myson

Contents

Plumbing and Heating

Plumbing systems

Plumbing is no longer the preserve of the professionals, thanks to a wide range of tools and fittings that are available to the DIY enthusiast. With the addition of easy-to-use plastic and metal fittings and pipework, successful repairs and modifications are easier – and faster – than ever.

The advantages of DIY plumbing

Being able to tackle your own plumbing installations and repairs can save you the cost of hiring professionals – and that can amount to a substantial sum of money. It also avoids the distress and inconvenience of ruined decorations, and the expense of replacing rotted household timbers where

a slow leak has gone undetected. Then there's the saving in water. A dripping tap wastes litres of water a day – and if it's hot water, there's the additional expense of heating it. A little of your time and a few pence spent on a washer can save you pounds.

Water systems

Generally, domestic plumbing incorporates two systems. One is the supply of fresh water from the 'mains', and the other is the waste or drainage system that disposes of dirty water. Both of the systems can be installed in different ways (see opposite).

flush sanitaryware during a temporary mains failure; the major part of the supply is under relatively low pressure, so the system is reasonably quiet; and because there are fewer mains outlets, there is less likelihood of impure water being siphoned back into the mains supply.

Stored-water system (Indirect)
The majority of homes are plumbed with a stored-water supply system. The storage tank in the loft and the cold-water tap in the kitchen are fed directly from the mains; so possibly are your washing machine, electric shower(s) and outside tap. But water for baths, washbasins, flushing WCs and some types of shower is drawn from the storage tank, which should be covered with a purpose-made lid to protect the water from contamination. Drinking water should only be taken from the cold-water tap in the kitchen.

Cold water from the storage tank is fed to a hot-water cylinder, where it is indirectly heated by a boiler or immersion heater to supply the hot taps. The water pressure at the taps depends on the distance ('head') from the tank to the tap.

A stored-water system provides several advantages. There is adequate water to

Mains-fed system (Direct)
Many properties now take all their water directly from the mains – all the taps are under high pressure, and they all provide water that's suitable for drinking. This development has come about as a result of limited loft space that precludes a storage tank and the introduction of non-return check valves, which prevent drinking water being contaminated. Hot water is supplied by a combination boiler or a multipoint heater; these instantaneous heaters are unable to maintain a constant flow of hot water if too many taps are running at once. Some systems use an unvented, pressurised cylinder, which stores hot water but is fed from the mains.

A mains-fed system is cheaper to install than an indirect one. Other advantages include mains pressure and drinking water at all taps. With a mains-fed system there's no plumbing in the loft to freeze.

Drainage

Waste water is drained in one of two ways. In houses built before the late 1950s, water is drained from baths, sinks and basins into a waste pipe that feeds into a trapped gully at ground level. Toilet waste feeds separately into a large-diameter vertical soil pipe that runs directly to the underground main drainage network.

With a single-stack waste system, which is installed in later buildings, all waste water drains into a single soil pipe – the one possible exception being the kitchen sink, which may drain into a gully.

Rainwater usually feeds into a separate drain, so that the house's drainage system will not be flooded in the event of a storm.

• Wiring Regulations
When making repairs or improvements to your plumbing, make sure you don't contravene the electrical Wiring Regulations. All metal plumbing has to be bonded to earth. If you replace a section of metal plumbing with plastic, it is important to reinstate the earth link. (See below.)

Reinstate the link
If you replace a section of metal plumbing with plastic, you may break the path to earth – so make sure you reinstate the link. Bridge a plastic joint in a metal pipe with an earth wire and two clamps. If you are in any doubt, consult a qualified electrician.

Water Regulations

The Water Supply (Water Fittings) Regulations 1999 (or Byelaws in Scotland) govern the design, installation and maintenance of plumbing systems, fittings and appliances that use water. These laws are intended to prevent the misuse, waste and contamination of water and apply from the point at which water enters your home's service pipe.

Owners, occupiers and anyone who installs plumbing or water fittings have a legal duty to ensure that systems meet the regulations and that water supplies aren't contaminated. They must also give advance notification to their local water supplier of various alterations to the system, including installation of a rim supply bidet; a bath with a capacity of more than 230 litres (50 gallons); shower pumps drawing more than 12 litres (2½ gallons) a minute; and garden watering systems.

Your local water supplier will provide you with the relevant information about other notifiable works, inspection requirements and possible certification for new work and for major alterations.

Drainage
The Building Regulations on drainage are designed to protect health and safety. Before undertaking work on your soil and waste pipes or drains (except for emergency unblocking) you need to contact the Building Control Office of your local authority.

SEE ALSO > Supplementary bonding 72, Earthing 77, Garden tap 50, Water bylaws 51

Direct and indirect systems

Stored-water system
Central heating omitted for clarity.

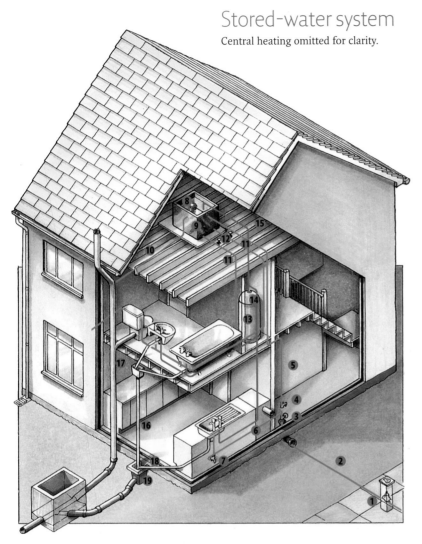

1 Water company stopcock
The water company uses this stopcock to turn off the supply to the house. Make sure it can be located quickly in an emergency.

2 Service pipe
From the water company stopcock onwards, the plumbing becomes the responsibility of the householder.

3 Household stopcock
The water supply to the house itself is shut off at this point.

4 Draincock
A draincock here allows you to drain water from the rising main.

5 Rising main
Mains-pressure water passes to the cold-water storage tank via the rising main.

6 Drinking water
Drinking water is drawn off the rising main to the kitchen sink.

7 Garden tap
The water company allows a garden tap to be supplied with mains pressure, provided it is fitted with a check valve.

8 Float valve
This valve shuts off the supply from the rising main when the cistern is full.

9 Cold-water storage tank
Stores from 230 to 360 litres (50 to 80 gallons) of water. Positioned in the roof, the tank provides sufficient 'head', or pressure, to feed the whole house.

10 Overflow pipe
Also known as a warning pipe, it prevents an overflow by draining water to the outside.

11 Cold-feed pipes
Water is drawn off to the bathroom and to the hot-water cylinder from the storage tank.

12 Cold-feed valves
Valves at these points allow you to drain the cold water in the feed pipe without having to drain the whole tank as well.

13 Hot-water cylinder
Water is heated and stored in this cylinder.

14 Hot-feed pipe
All hot water is fed from this point.

15 Vent pipe
Allows for expansion of heated water and enables air to be vented from the system.

16 Waste pipe
Surmounted by a hopper head, it collects water from basin and bath.

17 Soil pipe
Separate pipe takes toilet waste to main drains.

18 Kitchen waste pipe
Kitchen sink drains into same gully as waste pipe from upstairs.

19 Trapped gully

Mains-fed-water system
Central heating omitted for clarity.

1 Water-supplier's stopcock
May include water meter.

2 Service pipe

3 Main stopcock

4 Rising main
Supplies water directly to cold-water taps and WCs, etc.

5 Water heater or combination boiler

6 Unvented storage cylinder
(Not required for instantaneous heaters.)

7 Single-stack soil pipe
WC, handbasin, bath and shower drain into the stack. The stack may be fitted with an air-admittance valve terminating inside the house.

8 Sink waste
Water from the sink drains into a trapped gully.

9 Trapped gully

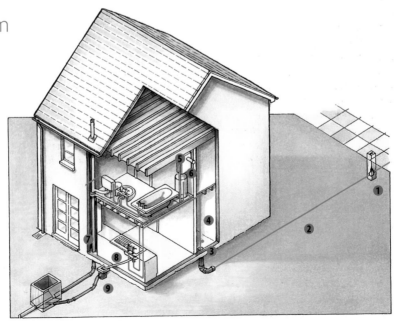

Water meters
Instead of paying a flat-rate water charge based upon the size of your home, you can opt to have your water consumption metered so you pay for what you use. For two people living in a large house, the savings can be considerable. Water meters are fitted to the incoming mains, usually outside at the supplier's stopcock, where they can be read more easily.

SEE ALSO > Wet central heating 55

Draining the system

You will have to drain at least part of any plumbing system before you can work on it; and if you detect a leak, you will have to drain the relevant section quickly. So find out where the valves, stopcock and draincocks are situated, before you're faced with an emergency.

Saving hot water
If your gate valve won't close off and you don't want to drain all the hot water, you can siphon the water out of the cold tank with a garden hosepipe. While the tank is empty, replace the old gate valve.

Draining cold-water taps and pipes

• Turn off the main stopcock on the rising main to cut off the supply to the kitchen tap (and to all the other cold taps on a direct system).
• Open the tap until the flow ceases.
• To isolate the bathroom taps, close the valve on the appropriate cold-feed pipe from the storage tank and open all taps on that section. If you can't find a valve, rest a wooden batten across the tank and tie the arm of the float valve to it. This will shut off the supply to the tank, so you can empty it by running all the cold taps in the bathroom. If you can't get into the loft, turn off the main stopcock, then run the cold taps.

Draining hot-water taps and pipes

• Turn off immersion heater or boiler.
• Close the valve on the cold-feed pipe to the cylinder and run the hot taps. Even when the water stops flowing, the cylinder will still be full.
• If there's no valve on the cold-feed pipe, tie up the float-valve arm, then turn on the cold taps in the bathroom to empty the storage tank. (If you run the hot taps first, the water in the tank will flush all the hot water from the cylinder.) When the cold taps run dry, open the hot taps. In an emergency, open hot and cold taps to clear the pipes as quickly as possible.

Draining a WC cistern

• To empty the WC cistern itself, tie up its float-valve arm and flush the WC.
• To empty the pipe that supplies the cistern, either turn off the main stopcock on a direct system or, on an indirect system, close the valve on the cold feed from the storage tank. Alternatively, shut off the supply to the storage tank and empty it through the cold taps. Flush the WC until no more water enters its cistern.

Draining the cold-water storage tank

• To drain the storage tank in the roof space, close the main stopcock on the rising main, then open all the cold taps in the bathroom. Bail out the residue of water at the bottom of the tank.

Closing a float valve
Cut off the supply of water to a storage tank by tying the float arm to a batten.

Draining the hot-water cylinder

• If the hot-water cylinder springs a leak (or you wish to replace it), first turn off the immersion heater and boiler, then shut off the cold feed to the cylinder from the storage tank (or drain the cold-water storage tank – see above). Run hot water from the taps.
• Locate a draincock from which you can drain the water remaining in the cylinder. It is probably located near the base of the cylinder, where the cold feed from the storage tank enters. Attach a hose and run it to a drain or sink that is lower than the cylinder. Turn the square-headed spindle on the draincock till you hear water flowing.
• Water can't be drained if the washer is baked onto the draincock seating, so disconnect the vent pipe and insert a hosepipe to siphon the cylinder.
• Should you want to replace the hot-water cylinder, don't disconnect all its pipework until you have drained the cylinder completely. If the water is heated indirectly by a heat-exchanger, there will be a coil of pipework inside the hot-water cylinder that is still full of water. This coil can be drained via the stopcock on the boiler after you have shut off the mains supply to the small feed-and-expansion tank in the roof space. Switch off the electricity to the central-heating system.

Adding extra valves

Unless you divide up the system into relatively short pipe runs with valves, you will have to drain off a substantial part of a typical plumbing installation even for a simple washer replacement.

• Install a gate valve on both the cold-feed pipes running from the cold-water storage tank. This will eliminate the necessity for draining off litres of water in order to isolate pipes and appliances on the low-pressure cold-and hot-water supply.
• When you are fitting new taps and appliances, take the opportunity to fit miniature valves on the supply pipes. In future, when you have to repair an individual tap or appliance, you will be able to isolate it in moments.

Gate valve
Fit a gate valve to the cold-feed pipes from the storage tank.

Miniature valve
Fit a miniature valve to the supply pipes below a sink or basin.

Sealed central-heating systems

A sealed system (see SEALED CENTRAL-HEATING SYSTEMS) does not have a feed-and-expansion tank – the radiators are filled from the mains via a flexible hose known as a filling loop. The indirect coil in the hot-water cylinder is drained as described left, though you might have to open a vent pipe that is fitted to the cylinder before the water will flow.

SEE ALSO > Cylinders 52, Cylinder vent pipe 52, Sealed central heating 55, Radiators 57, Consumer unit 76–77

Repairs and maintenance

It pays to master the simple techniques for coping with emergency repairs – in order to avoid the inevitable damage to your home and property, as well as the high cost of calling out a plumber at short notice. All you need is a simple tool kit and a few spare parts.

Draining and refilling the system

Partially drain the plumbing system if you are leaving the house unoccupied for a few days during winter and leave the central heating on a low setting. For longer periods away at any time of the year, it's wise to drain the system completely.

Attach hosepipe to draincock

Partial drain-down
• Add special antifreeze to the central-heating feed tank and set the heating to come on briefly twice a day.
• Turn off the main stopcock.
• Open all taps to drain the system.

Full drain-down
• Switch off and extinguish the water heater and/or boiler.
• Turn off the main stopcock and, if possible, the main stopcock outside.
• Open all taps to drain the pipework.
• Open the draincock at the base of the hot-water cylinder. If there are draincocks in the rising main or other pipework, open them too.
• Flush the WCs.
• Drain the boiler and radiator circuits at the lowest points on the pipe runs.
• Add salt to the WC pan to prevent the trap water freezing.

Refilling the system
• Close all taps and draincocks.
• Turn on the main stopcock.
• Turn on taps and allow water and air to escape. Bleed radiators and check that float valves are working properly.

Thawing frozen pipes

If water won't flow from a tap during cold weather, or a tank refuses to fill, a plug of ice may have formed in one of the supply pipes. The plug cannot be in a pipe supplying taps or float valves that are working normally, so you should be able to trace the blockage quickly. In fact, freezing usually occurs first in the roof space.

As copper pipework transmits heat quickly, use a hairdryer to gently warm the suspect pipe, starting as close as possible to the affected tap or valve and working along it. Leave the tap open, so water can flow normally as soon as the ice thaws. If you can't heat the pipe with a hairdryer, wrap it in a hot towel or hang a hot-water bottle over it.

Preventive measures
Insulate pipework and fittings to stop them freezing, particularly those in the loft or under the floor. If you're going to leave the house unheated for a long time during the winter, drain the system (see left). Cure any dripping taps, so leaking water doesn't freeze in your drainage system overnight.

Thawing a frozen pipe
Play a hairdryer gently along a frozen pipe, working away from the blocked tap or valve.

Dealing with a punctured pipe

Unless you are absolutely sure where your pipes run, it is all too easy to nail through one of them when fixing a loose floorboard. You may be able to detect a hissing sound as water escapes under pressure, but more than likely you won't notice your mistake until a wet patch appears on the ceiling below, or some problem associated with damp occurs at a later date. While the nail is in place, water will leak relatively slowly, so don't pull it out until you have drained the pipework and can repair the leak. If you pull out the nail by lifting a floorboard, replace the nail immediately.

If you plan to lay fitted carpet, you can paint pipe runs on the floorboards to avoid such accidents in future.

Patching a leak

During freezing conditions, water within a pipe turns to ice, which expands until it eventually splits the walls of the pipe or forces a joint apart. The only other reason for leaking plumbing is mechanical failure – either through deterioration of the materials or because a joint has failed and is no longer completely watertight. Make a permanent repair if you can by inserting a new section of pipe or replacing a leaking joint (if it is a compression joint that has failed, try tightening it first). If you have to make an emergency repair, drain the pipe first. If it is frozen, make the repair before it thaws.

Sealing a split or pinhole
For a temporary repair, use a repair clamp, below, or special amalgamating tape which will work in wet conditions.

A push-fit repair pipe (see right), makes a fast, permanent repair if you have to remove a section of pipe.

Alternatively, a slip-on compression fitting (below left) will also make a permanent repair.

A push-fit repair pipe

A slip-on compression fitting makes a secure repair

Fit an emergency repair clamp over the split

SEE ALSO >Joining pipes 21–7, Compression joints 22

Repairing leaking taps

Leaking taps aren't too difficult to deal with. When water drips from a spout, it usually points to a faulty washer. If the tap is old, its seat may be worn, too. If water leaks from beneath the head of the tap when it's in use, the O-ring needs replacing. When you are working on a tap, insert the plug and lay a towel in the bottom of the washbasin, bath or sink to catch small objects and protect the surface.

Replacing a washer

Traditional pillar tap
The components of a pillar tap

1 Capstan head	nut
2 Metal shroud	**6** Jumper
3 Gland nut	**7** Washer
4 Spindle	**8** Tap body
5 Headgear	**9** Seat
	10 Tail

To replace the washer in a traditional pillar tap, first drain the supply pipe, then open the valve as far as possible.

If the tap is shrouded with a metal cover, unscrew it by hand or use a wrench, taping the jaws to protect the chrome finish.

Lift the cover to reveal the headgear nut just above the body of the tap. Slip a spanner onto the nut and unscrew it (**1**).

The jumper to which the washer is fixed fits into the bottom of the headgear. In some taps the jumper is removed along with the headgear (**2**), but in other types it will be lying inside the tap body.

The washer itself may be pressed over a small button in the centre of the jumper (**3**) and can be prised off with a screwdriver. A securing nut can be hard to remove. Allow penetrating oil to soften any corrosion; then, hold the jumper stem with pliers and unscrew the nut with a spanner (**4**). If the nut sticks, replace the jumper and washer; otherwise, fit a new washer and retaining nut, then reassemble the tap.

1 Loosen headgear nut

3 Prise off washer

2 Lift out headgear

4 Or undo securing nut

Curing a dripping ceramic-disc tap

1 Prise out the disc

2 Pull the head off

Getting inside a tap
On most modern taps the head and cover is in one piece. You will have to remove it to expose the headgear nut. Often a retaining screw is hidden beneath the coloured hot/cold disc in the centre of the head. Prise out the disc with the point of a knife or a small screwdriver (**1**). If there's no retaining screw, simply pull the head off (**2**).

1 Unscrew and examine the valve

3 Unscrew retaining nut

2 Replace a worn rubber seal

4 Replace inlet seals

Ceramic-disc taps should be maintenance free, but faults can still occur. Since there's no washer to replace, you have to replace the whole valve when the tap leaks.

Turn off the water. Remove the headgear by turning it anticlockwise with a spanner. Remove the valve and examine it for wear or damage (**1**). Removing debris from the disc may do, but if disc is scratched, you will need a new valve. These are handed – left (hot), right (cold). Also examine the rubber seal on the bottom of the valve. If worn or damaged, it will cause the tap to drip. If need be, replace the seal with a new one (**2**).

Single-lever mixer taps have a cartridge that controls both flow and temperature. To replace it, remove handle cap. Use a socket spanner to undo the handle nut. Lift handle off, remove shroud and undo the cartridge-retaining nut (**3**), then lift out the cartridge. Fit new inlet seals on underside of cartridge if discs appear undamaged (**4**); replace the entire cartridge if they are visibly damaged.

SEE ALSO > Bib tap 20, Tap mechanisms 30

Repairing seats and glands

If replacing a washer doesn't solve the problem then the tap itself is probably worn. If you want to keep the taps, it is possible to renovate them.

Regrinding or renewing the seat

If a tap continues to drip after you have replaced the washer, the seat is probably worn, allowing water to leak past the washer. A simple way to cure this is to cover the old seat with a nylon liner that is sold with a matching jumper and washer (1). Drop the liner over the old seat, replace the jumper and assemble the tap. Close the tap to force the liner into position.

Alternatively, use a tap grinding tool (2) to put a new smooth surface on the seat. The tool uses serrated cutters to grind out any imperfections that might be allowing water past the washer.

With the headgear removed, insert the grinder (3) and screw the threaded bush to the tap to locate the tool securely.

Push down and twist the grinder clockwise. Resurfacing is complete when the whole seat is shiny and new looking.

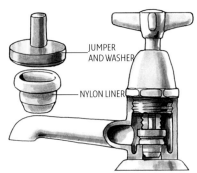

1 Repair a worn seat with a nylon liner

2 A tap grinding tool   3 Cut new seat

Stopcocks and valves

Stopcocks and gate valves are used rarely but often fail just when needed.

Make sure that they are operating smoothly by closing and opening them from time to time. Lubricate stiff spindles with penetrating oil. A stopcock is fitted with a standard washer, but as it is hardly ever under pressure it is unlikely to wear.

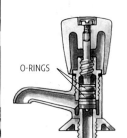

Penetrating oil or lubricant prevents seizure

Replacing O-rings

On a mixer tap each valve is usually fitted with a washer, as on conventional taps, but in most mixers the gland packing (see far right) has been replaced by a rubber O-ring. The base of a mixer's swivel spout is also sealed with a washer or O-ring. If water seeps from that, it needs replacing.

First remove the mixer spout – the retaining screw may be accessible from the front; if it's at the back, you may have to use a cranked screwdriver to remove it (1). Lift out the swivelling spout (2), then use a small screwdriver to prise out the O-ring from its groove (3). Take care not to scratch the metal. Lubricate a new ring with silicone grease (4), then carefully slide it into place before re-fitting the spout (5) and replacing the retaining screw.

1 Use a cranked screwdriver for rear screws
3 Remove the O-ring from the swivel spout
5 Replace spout

2 Lift out the swivelling spout
4 Lubricate a new O-ring

GLAND PACKING

Gland packing
Older-style taps are sealed with water-tight packing around the spindle.

O-RINGS

O-ring seal
Modern taps are sealed with rubber rings, in place of gland packing.

SEE ALSO > Gate valve 20, PTFE tape 22, Stopcock 24, Tap mechanisms 30

Maintaining cisterns and storage tanks

The mechanisms used in WC cisterns and storage tanks are probably the most overworked of all plumbing components, so servicing is required from time to time to keep them operating properly. You can get the spare parts you need from plumbers' merchants and DIY stores.

Direct-action WC cisterns

Most modern WCs are washed down by means of direct-action cisterns. Water enters the cistern through a valve, which is opened and closed by the action of a hollow float attached to one end of a rigid arm. As the water rises in the cistern, it lifts the float until the other end of the arm closes the valve and shuts off the supply.

Flushing is carried out by depressing a lever, which is linked by wire to a rod attached to a perforated plastic or metal plate at the bottom of an inverted U-bend tube (siphon). As the plate rises, the perforations are sealed by a flexible plastic diaphragm (flap valve), so the plate can displace a body of water over the U-bend to promote a siphoning action. The water pressure behind the diaphragm lifts it, so that the contents of the cistern flow up through the perforations in the plate, over the U-bend and down the flush pipe. As the water level in the cistern drops, so does the float – thus opening the float valve to refill the cistern.

Servicing cisterns

The few problems associated with this type of cistern are easy to solve. A faulty float valve or poorly adjusted float arm will allow water to leak into the cistern until it drips from the overflow pipe that runs to the outside of the house. Slow or noisy filling can often be rectified by replacing the float valve. If the cistern will not flush until the lever is operated several times, the flap valve probably needs replacing (see below). If the flushing lever feels slack, check that the wire link at the end of the flushing arm is intact. When water runs continuously into the pan, check the condition of the washer at the base of the siphon.

Direct-action cistern
The components of a typical direct-action WC cistern.
1 Float valve
2 Float
3 Float arm
4 Flushing lever
5 Wire link
6 Perforated plate
7 One-piece siphon
8 Flap valve
9 Overflow
10 Sealing washer
11 Retaining nut
12 Flush-pipe connector

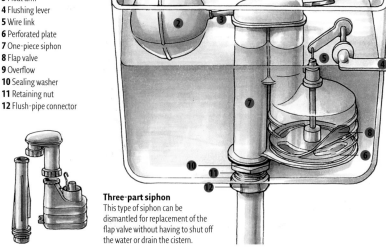

Three-part siphon
This type of siphon can be dismantled for replacement of the flap valve without having to shut off the water or drain the cistern.

Miniature float valve
This type of float valve is designed for installing in WC cisterns only.

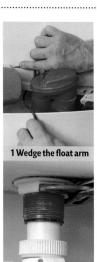

1 Wedge the float arm

2 Release flush pipe

3 Loosen retaining nut

4 Lift off flap valve

Replacing a flap valve

If a WC cistern will not flush first time, take off the lid and check that the lever is actually operating the flushing mechanism. If that appears to be working normally, then try replacing the flap valve in the siphon. Before you service a one-piece siphon, shut off the water by wedging the float arm with a bent wire across the cistern (**1**). Flush the cistern.

Use a wrench to unscrew the nut that holds the flush pipe to the underside of the cistern (**2**). Move the pipe to one side.

Release the retaining nut that clamps the siphon to the base of the cistern (**3**). A little water will run out as you loosen the nut – so have a bucket handy. You may find that the siphon is bolted to the base of the cistern, instead of being clamped by a single retaining nut.

Disconnect the flushing arm, then ease the siphon out of the cistern. Lift the diaphragm off the plastic plate (**4**) and replace it with one of the same size. Reassemble the entire flushing mechanism and reconnect the flush pipe to the cistern.

Making a new wire link

You won't be able to flush a WC cistern if the flushing lever has come adrift.

Make a replacement for a rusted link from thick wire; if the lever connecting the handle to the link has broken, the WC can be flushed by pulling the wire link upwards until you can get a new one.

Curing continuous running water

If you notice that water is running into the pan continuously, turn off the supply and let the cistern drain. If the siphon hasn't split, try changing the sealing washer.

If the water is flowing from the float valve so quickly that the siphoning action is not interrupted, fit a float-valve seat with a smaller water inlet.

SEE ALSO > Adjusting the float arm 14

Diaphragm valves

The pivoting end of the float arm on a diaphragm valve (known in the trade as a Part 2 valve) presses against the end of a small plastic piston, which moves the large rubber diaphragm to seal the water inlet.

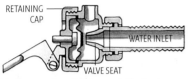

RETAINING CAP
WATER INLET
VALVE SEAT

Diaphragm valve: retaining cap to the front

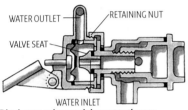

WATER OUTLET
RETAINING NUT
VALVE SEAT
WATER INLET

Diaphragm valve: retaining nut to the rear

Replacing the diaphragm

Turn off the water supply, then unscrew the large retaining cap. Depending on the model, the nut may be screwed onto the end of the valve or behind it (see above).

With the latter type of valve, slide out the cartridge inside the body (1) to find the diaphragm behind it. With the former, you will find a similar piston and diaphragm immediately behind the retaining cap (2).

Wash the valve, before assembling it along with the new diaphragm.

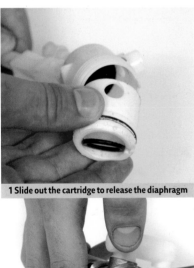

1 Slide out the cartridge to release the diaphragm

2 Undo the cap and pull float arm to find the valve

Close-coupled WCs and push-button cisterns

This modern style of WC does away with the pipe that connects the cistern to the pan. Water flows from the cistern through a moulded-in channel in the pan to the rim. The cistern is bolted directly to the pan and a rubber seal prevents leaks at the joint. This style often features compact cisterns with push-button action and lower water usage.

The cistern may incorporate a diaphragm valve, but it's becoming more common to find close-coupled WCs with push-button, 'continental'-style cisterns. Here, the traditional handle is replaced by a button on the top of the cistern. Usually the button is split, to give a dual flush facility of 4 or 6 litres (7 or 10½ pints), helping to save water.

Push-button cisterns use a different type of valve, which is activated via a cable release. A small float controls the level of water in the cistern. The valve usually incorporates an overflow, which allows water to flow into the toilet bowl – rather than outside via an overflow pipe – if there is a problem.

The mechanism inside the cistern is all plastic, so won't rust. As a result, there's less to go wrong, but if problems occur spare parts may be harder to find as there are many different types available.

These cisterns produce a faster flush, which doesn't give the same cleansing action as a slower, syphon-operated flush. However, they are quiet in operation and allow for a very slimline cistern, helping to reduce the 'footprint' of the WC.

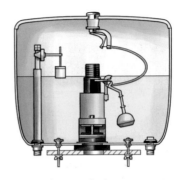

Push-button cistern mechanism

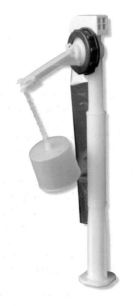

A close-coupled cistern with push-button operation

Flapper valve
This is a modern replacement for the siphon and can be operated by handle or push-button. It helps to save water because it only allows water to flush while the handle or button is depressed, unlike a siphon, which empties a full cistern every time. It has a built-in overflow, but as it lies in the bottom of the cistern it can be prone to limescale or debris interfering with the seal. It can be used in conjunction with all types of float valve.

Other types of cistern valve

A Fluidmaster valve converts a lever-operated siphon into a push-button cistern. A flap valve controls the flow of water into the WC. Quiet in use and water-efficient, it works well with low-capacity cisterns, but is not suitable for double-trap toilets. Two kits are available: one which replaces the lever and siphon; the other which also replaces the float valve.

The Torbeck valve can be used to replace the diaphragm or piston valve in most cisterns. Its all-plastic construction means there's nothing to corrode. After flushing, water enters the cistern through a tube which is mostly below water level, so there's no noise from splashing. The float is easily adjustable via a screw thread. Side and bottom entry versions are available, the latter with a built-in debris filter and flow control device.

Fluidmaster valve

Torbeck valve

SEE ALSO > Turning off the water 8, Adjusting the float arm 14

Renovating valves and floats

A cistern or tank that doesn't work properly can prevent WCs flushing properly or cause noise throughout the plumbing system. Sorting the problem can be as simple as bending the float arm or adjusting a screw.

Adjusting the float arm

Adjust the float so as to maintain the optimum level of water, which is about 25mm (1in) below the outlet of the overflow pipe.

Bend the metal arm on a Portsmouth valve downward to reduce the water level, or straighten it to let in more water (**1**).

The arm on a diaphragm valve has an adjusting screw, which raises or lowers the arm to alter the water level (**2**).

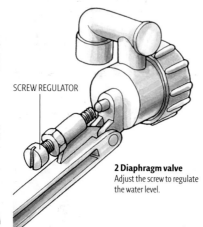

1 Straighten or bend a metal float arm

Thumb-screw adjustment
Some float arms are cranked, and the float is attached with a thumb-screw clamp. To adjust the water level in the cistern, slide the float up or down the rod.

Float valve with flexible silencer tube

SCREW REGULATOR

2 Diaphragm valve
Adjust the screw to regulate the water level.

Replacing the float

Modern plastic floats rarely leak, but old-style metal floats eventually corrode and allow water to seep into the ball. The float gradually sinks, until it won't ride high enough to close the valve. Unscrew the float to see if there is water inside.

If you can't obtain a new float for several days, lay the ball on a bench, enlarge the leaking hole with a screwdriver and pour out the water. Cover the ball with a plastic bag, tying the neck tightly around the float arm, and then replace the float.

Curing noisy cisterns

Cisterns that fill noisily can be very annoying. It was once permitted to screw a pipe into the outlet of a valve so that it hung vertically below the level of the water. This solved the splashing problem, but concern about the possibility of water 'back-siphoning' through the silencer tube into the mains supply led to rigid tubes being banned in favour of flexible plastic silencer tubes (see far left), which seal by collapsing should back-siphoning occur.

A silencer tube can also prevent water hammer – a rhythmic thudding that reverberates along the pipework. This is often the result of ripples on the surface of the water in a cistern, caused by a heavy flow from the float valve. As the water rises, the float arm bouncing on the ripples 'hammers' the valve, and the sound is amplified and transmitted along the pipes. A flexible plastic tube will stop ripples by introducing water below the surface.

If the pressure through the valve is too high, the arm oscillates as it tries to close the valve – causing water hammer. This can be cured by fitting an equilibrium valve. As water flows through the valve, some of it is introduced behind the piston or diaphragm to equalize the pressure on each side, so that the valve closes smoothly and silently.

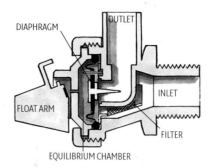

DIAPHRAGM   OUTLET   INLET   FILTER
FLOAT ARM   EQUILIBRIUM CHAMBER

Diaphragm-type equilibrium valve

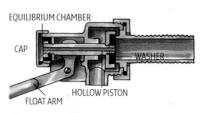

EQUILIBRIUM CHAMBER   CAP   WASHER   HOLLOW PISTON   FLOAT ARM

Piston-type equilibrium valve

Renewing a float valve

Turn off the supply of water to the cistern or tank, then use a spanner to loosen the tap connector joining the supply pipe to the float-valve stem. Remove the float arm, then unscrew the fixing nut outside of the cistern and pull out the valve.

Fit the replacement valve and, if possible, use the same tap connector to join it to the supply pipe.

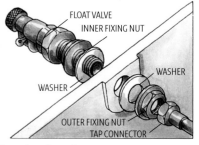

FLOAT VALVE
INNER FIXING NUT
WASHER
WASHER
OUTER FIXING NUT
TAP CONNECTOR

Renewing a float valve
Clamp the valve to the cistern with fixing nuts.

Choosing the correct pressure

Float valves are made to suit different water pressures: low, medium and high (LP, MP and HP). If the pressure is too low for the valve, the cistern may take a long time to fill; if too high, it may leak continuously. Most domestic WC cisterns require an LP valve; those fed direct from the mains need an HP valve. If the head (the height of the tank above the float valve) is greater than 13.5m (45ft), fit an MP valve; if over 30m (100ft), fit an HP valve. In an apartment with a packaged plumbing system (a storage tank built on top of the hot-water cylinder), the pressure may be so low that you will have to fit a full-way valve to the WC cistern in order to get it to fill reasonably quickly. If you live in an area where water pressure fluctuates a great deal, fit an equilibrium valve (see left).

To alter the pressure of a modern valve, simply replace the seat inside.

SEE ALSO > Float valves 13, Supporting pipes 28

Drainage systems

A drainage system is designed to carry dirty water and WC waste from the appliances in your home to underground drains leading to the main sewer. The various branches of the waste system are protected by U-bend traps full of water, to stop drain smells fouling the house. Depending on the age of your house, it will have a two-pipe system or a single stack. Because the two-pipe system has been in use for very much longer, it is still the more common of the two. Use similar methods to maintain either system.

Responsibility for the drains

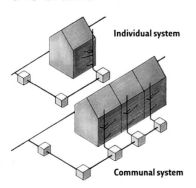

Individual system

Communal system

If a house is drained individually, the whole system up to the point where it joins the sewer is the responsibility of the householder. However, where a house is connected to a communal drainage system linking several houses, the arrangement for maintenance, including the clearance of blockages, is not so straightforward.

If the drains were constructed prior to 1937, the local council is responsible for cleansing but can reclaim the cost of repairing any part of the communal system from the householders. After that date, all responsibility falls upon the householders collectively, so that they are required to share the cost of the repair and cleansing of the drains up to the sewer, no matter where the problem occurs. Contact the Technical Services Department of your local council to find out who is responsible for your drains.

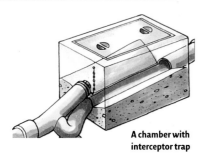

A chamber with interceptor trap

Two-pipe system

The waste pipes of older houses are divided into two separate systems. WC waste is fed into a vertical soil pipe that leads directly to the underground drains. To discharge drain gases at a safe height and make sure that back-siphoning cannot empty the WC traps, the soil pipe is vented to the open air above the guttering.

Individual branch pipes leading from upstairs washbasins and baths drain into an open hopper that funnels the water into another vertical waste pipe. Instead of feeding directly into the underground drains, this pipe terminates over a yard gully – another trap covered by a grid. A separate waste pipe from the kitchen sink normally drains into the same gully.

The yard gully and soil pipe both discharge into an underground inspection chamber, or manhole. These chambers provide access to the main drains for clearing blockages, and are located wherever the main drain changes direction on its way to the sewer.

At the last inspection chamber, just before the drain enters the sewer, there is an interceptor trap, the final barrier to drain gases and sewer rats.

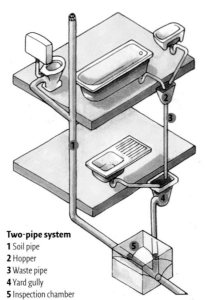

Two-pipe system
1 Soil pipe
2 Hopper
3 Waste pipe
4 Yard gully
5 Inspection chamber

Single-stack system

Since the late 1950s, most houses have been drained using a single-stack system. Waste from basins, baths and WCs is fed into the same vertical soil pipe or stack – which, unlike the two-pipe system, is often built inside the house. A single-stack system must be designed carefully to prevent a heavy discharge of waste from one appliance siphoning the trap of another, and to avoid the possibility of WC waste blocking other branch pipes. The vent pipe of the stack terminates above the roof and is capped with an open cage; or inside the house and is fitted with an air-admittance valve (see right).

The kitchen sink can be drained through the same stack, but it is still common practice to drain sink waste into a yard gully. Nowadays waste pipes must pass through the grid, stopping short of the water in the gully trap – so that if blocked with leaves, the waste can discharge unobstructed into the gully. Alternatively, it may be a back-inlet gully, with the waste pipe entering below ground level.

A downstairs WC is sometimes drained through its own branch drain to an inspection chamber.

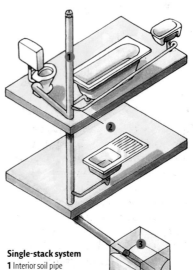

Single-stack system
1 Interior soil pipe
2 All branch pipes run to stack
3 Inspection chamber

Ventilating pipes and stacks
An air-admittance valve seals off the vent pipe, but allows air into the system to prevent water being siphoned from the trap seals. This type of valve can only be used if the drainage scheme has been approved by the local authority.

Prefabricated chamber
On a modern drainage system, the inspection chambers may take the form of cylindrical prefabricated units. There may not be an interceptor trap in the last chamber before the sewer.

SEE ALSO > Plumbing systems 6–7, Blocked soil pipe 17, Yard gully 17, Blocked drains 18

Clearing blockages

Don't ignore the early signs of an imminent blockage in the waste pipe from a sink, bath or basin. If the water drains away slowly, use a chemical cleaner to remove a partial blockage before it becomes more serious. If a waste pipe blocks without warning, there are various ways to locate and clear the obstruction.

Cleansing the waste pipe

Grease, hair and particles of kitchen debris build up gradually within the traps and waste pipes. Regular cleaning with a proprietary chemical drain cleaner will keep the waste system clear and sweet-smelling.

If water drains away sluggishly, use a cleaner immediately. Follow the manufacturer's instructions carefully, with particular regard to safety. Always wear protective gloves and goggles when handling chemical cleaners, and keep them out of the reach of children.

If unpleasant odours linger after you've cleaned the waste, pour a little disinfectant into the basin overflow.

Use a plunger to force out a blockage

Using a plunger or pump

If one basin fails to empty while others are functioning normally, the blockage must be somewhere along its individual branch pipe. Before you attempt to locate the blockage, try forcing it out of the pipe with a sink plunger. Smear the rim of the rubber cup with petroleum jelly, then lower it into the blocked basin to cover the waste outlet. Make sure that there's enough water in the basin to cover the cup. Hold a wet cloth in the overflow with one hand and pump the handle of the plunger up and down a few times. The waste may not clear immediately if the blockage is merely forced further along the pipe, so repeat until the water drains away.

If it will not clear after several attempts, try hiring a hand-operated hydraulic pump to clear the pipe. Block the sink overflow with a wet cloth and fill the pump with water from the tap. Hold the pump's nozzle over the outlet, pressing down firmly, then pump up and down until the obstruction is cleared.

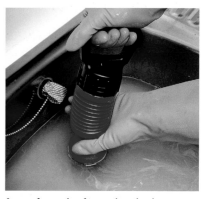
A pump forces a jet of water along the pipe

Clearing the trap

If you can't shift the blockage using a plunger or pump, dismantle the trap situated immediately below the waste outlet of a sink or basin. The trap is basically a bent tube designed to hold water to seal out drain odours. Traps become blocked when debris collects at the lowest point of the bend.

Place a bucket under the basin to catch the water, then use a wrench to release the cleaning eye at the base of a P-trap; on a bottle trap, remove the large access cap by hand. If there is no provision for gaining access to the trap, unscrew the connecting nuts and remove it.

Let the contents of the trap drain into the bucket, then bend a hook on the end of a length of wire and use it to probe the section of waste pipe beyond the trap. (It is also worth checking outside, to see if the other end of the pipe is blocked with leaves.) If you have had to remove the trap, take the opportunity to scrub it out with detergent before replacing it.

Cleaning the branch pipe

Quite often, a vertical pipe from the trap joins a virtually horizontal section of the waste pipe. Unscrew the access plug built into the joint and use a length of hooked wire to probe the branch pipe. If you locate a blockage that seems very firmly lodged, rent a drain auger from a tool-hire company to clear the pipework.

If there's no access plug, remove the trap and probe the waste pipe with an auger. If it's made with push-fit joints, it can be dismantled easily.

Undo the cleaning eye on a P-trap

Unscrew the access cap on a bottle trap

Use an auger to clear a branch pipe

SEE ALSO > Frozen pipes 9

Blocked stacks or gullies

If several fittings are draining poorly, the vertical stack is probably obstructed. In autumn, the hopper, downpipe and yard gully may be blocked with leaves. The blockage may not be obvious when you empty a basin, but the contents of a bath will almost certainly cause an overflow. Clear the blockage immediately to avoid penetrating damp affecting your house.

Cleaning out the hopper and drainpipe

Wearing protective gloves, scoop out the debris from the hopper, then gently probe the drainpipe with a cane to check that it is free. Clear the bottom end of the pipe with a piece of bent wire. If an old cast-iron waste pipe has been replaced with a modern plastic pipe, you may find there are cleaning eyes or access plugs at strategic points for clearing a blockage.

While you're on the ladder, scrub the inside of the hopper and disinfect it to prevent odours entering the bathroom.

Unblocking a yard gully

Unless you decide to hire an auger, you have little option but to clear a blocked gully by hand. However, by the time it overflows the water in the gully will be quite deep, so try bailing some of it out with a small disposable container. Wearing rubber gloves, scoop out the debris from the trap until the remaining water disperses.

Rinse the gully with a hose and cleanse it with disinfectant. Scrub the grid as clean as possible, or burn off accumulated grime from a metal grid with a gas torch.

If a flooded gully appears to be clear and yet the water will not drain away, try to locate the blockage at the nearest inspection chamber.

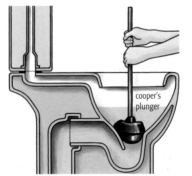
Bail out the water, then clear a gully by hand

Unblocking a soil pipe

Unblocking a soil pipe is an unpleasant job, which can be doubly difficult because of access problems. A cast-iron external stack will almost certainly have to be cleared via the vent above the roof, while a stack running inside the house may be hard to reach and any mess could ruin carpets or wallpaper. In either case it may be best to call in a professional cleaning company.

You can clean an external plastic stack yourself, since there should be a large access plug or cleaning eye wherever branch pipes join the stack. Unscrew and open the cleaning eye to insert a hired drain auger. Pass the auger into the stack until you locate the obstruction, then crank the handle to engage it. Push or pull the auger until you can dislodge the obstruction to clear the trapped water, then hose out the stack. Wash and disinfect the surrounding area.

Use a hired auger to clear a soil stack

Unblocking a WC

If the water in a WC pan rises when you flush it, there's a blockage in the vicinity of the trap. A partial blockage allows the water level to fall slowly.

Hire a larger version of the sink plunger to force the obstruction into the soil pipe. Position the rubber cup of the plunger well down into the U-bend, and pump the handle. When the blockage clears, the water level will drop suddenly, accompanied by an audible gurgling.

If the trap is blocked solidly, hire a special WC auger. Pass the flexible clearing rod as far as possible into the trap, then crank the handle to dislodge the blockage. Wash the auger in hot water and disinfect it, before returning it to the hire company. A more serious blockage may require professional assistance.

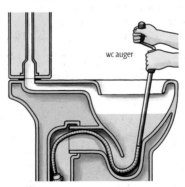

• **Clearing a blockage with a hydraulic pump**
Shift a really stubborn blockage with a hired pump, similar to the one used for clearing a blocked sink (see opposite).

cooper's plunger

wc auger

Clearing a blockage
Use a Cooper's plunger (top) to pump a blocked WC. Alternatively, clear it with a special WC auger (above).

SEE ALSO > Inspection chambers 18

Rodding the drains

The first sign of a blocked drain may be an unpleasant smell from an inspection chamber, but a severe blockage may cause waste to overflow from a gully or inspection chamber. Before calling in the professionals, hire a set of drain rods – flexible rods made of plastic or wire, screwed end to end – to clear the blockage.

Locating the blockage

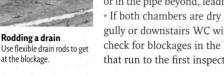

Rodding a drain
Use flexible drain rods to get at the blockage.

Lift the cover from the inspection chamber nearest to the house. If it's stuck or the handles have rusted away, scrape the dirt from around its edges and prise it up with a garden spade.
• If the chamber contains water, check the one nearer the road or boundary. If that chamber is dry, the blockage is between the two chambers.

If the chamber nearest the road is full, the blockage will be in the interceptor trap or in the pipe beyond, leading to the sewer.
• If both chambers are dry and a yard gully or downstairs WC will not empty, check for blockages in the branch drains that run to the first inspection chamber.

Rodding the drainpipe

Screw two or three rods together and attach a corkscrew fitting to the end. Insert the rods into the drain at the bottom of the inspection chamber, in the direction of the suspected blockage. If the chamber is full of water, use the end of a rod to locate the open channel running across the floor, leading to the drain opening.

As you pass the rods along the pipe, attach further lengths till you reach the obstruction, then twist the rods clockwise to engage the screw (don't twist the rods anticlockwise, or they will become detached). Agitate the obstruction until it breaks up, allowing the water to flow away.

Extract the rods, flush the chamber with clean water from a hose, and then replace the cover and clean the rods.

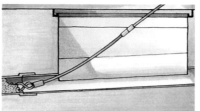

Use a corkscrew fitting to clear a drain

Clearing traps

Screw a rubber plunger to the end of a short length of rods and locate the channel that leads to the base of the trap. Push the plunger into the opening of the trap, then pump the rods a few times to expel the blockage. (This is also a useful technique for clearing blocked yard gullies.)

If the water level does not drop after several attempts, try clearing the drain leading to the sewer. Access to this drain is through a cleaning eye above the trap. It will be sealed with a stopper, which you will have to dislodge with a drain rod, unless it is attached to a chain stapled to the chamber wall. Don't let the stopper fall into the channel and block the trap. Rod the drain, then hose out the chamber before replacing the stopper and cover.

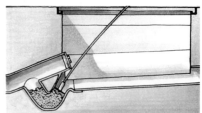

To rod an interceptor trap, fit a rubber plunger

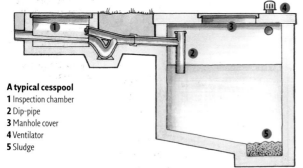

A typical cesspool
1 Inspection chamber
2 Dip-pipe
3 Manhole cover
4 Ventilator
5 Sludge

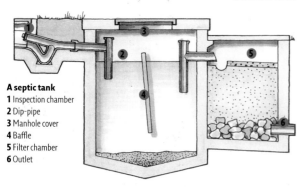

A septic tank
1 Inspection chamber
2 Dip-pipe
3 Manhole cover
4 Baffle
5 Filter chamber
6 Outlet

Cesspools and septic tanks

Houses built in the country or on the outskirts of a town may not be connected to a public sewer. Instead, waste is drained into a cesspool or septic tank.

A cesspool simply acts as a collection point for sewage until it can be pumped out by the local council. A septic tank is a complete waste-disposal system, in which sewage is broken down by bacterial action before the water is finally discharged into a local waterway or distributed underground.

Cesspools
The Building Regulations stipulate that cesspools must have a minimum capacity of 18,000 litres (4,000 gallons), but many existing cesspools hold far less and require emptying perhaps once every two weeks. Water boards estimate usage at around 115 litres (25 gallons) per person per day.

Septic tanks
The sewage in a septic tank separates slowly: heavy sludge falls to the bottom to leave relatively clear water, with a layer of scum floating on the surface. Waste discharges below the surface, so that incoming water does not stir up the sewage. Bacterial action takes a minimum of 24 hours, so the tank has baffles to slow down the movement of sewage through it.

The partly treated waste passes out of the tank into some form of filtration system for further bacterial action. This may consist of another chamber, containing a deep filter bed; or the waste may flow underground through a network of drains, which disperses the water over a wide area to filter through the soil.

Modern packaged waste water treatment plants are electrically operated and produce very clean effluent.

Rodding points

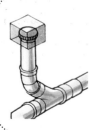

A modern drainage system is often fitted with rodding points to provide access to the drain. They are sealed with small oval or circular covers that may be a push or screw fit.

SEE ALSO > Inspection chambers 15

Metal pipes

The ability to install a run of pipework, make watertight joints and connect up to fittings constitutes the basis of most plumbing. Without these skills, a householder is restricted to simple maintenance. Modern materials and technology have made it possible for anybody who is prepared to master a few techniques to upgrade and extend plumbing without having to hire a professional.

Metric and imperial pipes

Copper and stainless-steel pipes are now made in metric sizes, whereas pipework already installed in older houses will have been made to imperial measurements. If you compare the equivalent dimensions (15mm – ½in, 22mm – ¾in, 28mm – 1in), the difference seems obvious, but metric pipe is measured externally while imperial pipe is measured internally. In fact, the difference is very small – but enough to cause some problems when joining one type of pipe to the other.

When making soldered joints, an exact fit is essential. Imperial to metric adaptors are necessary when joining 22mm pipe to its imperial equivalent; and, although not essential, adaptors are convenient when you are working with 28mm pipes or with thick-walled ½in pipes. Adaptors are not required when using compression fittings, but when you are connecting 22mm to ¾in plumbing slip an imperial olive onto the ¾in pipe.

Typically, 15mm (½in) pipe is used for the supply to basins, kitchen sinks, washing machines, some showers, and radiator flow and returns. However, 22mm (¾in) pipes are used to supply baths, high-output showers, hot-water cylinders and main central-heating circuits; and 28mm (1in) pipe for larger heating installations.

Electrochemical action

Joining pipes made from different metals can accelerate corrosion as a result of electrolytic action. If you live in a soft-water area, where this problem tends to be pronounced, use plastic pipe and connectors when you're joining to old pipework – but make sure that the metal pipes are still bonded to earth, as required by the Wiring Regulations.

Metal supply pipes

Over the years, most household plumbing systems will have undergone some form of improvement or alteration. As a result, you may find any of a number of metals used, perhaps in combination, depending on the availability of materials at the time of installation or the preference of an individual plumber.

Copper

Half-hard-tempered copper tubing is the most widely used material for pipework. This is because it's lightweight, solders well, and can be bent easily (even by hand, with the aid of a bending spring). It is used for both hot-water and cold-water pipes, as well as for central-heating systems. There are three sizes of pipe generally used for domestic plumbing: 15mm (½in), 22mm (¾in), and 28mm (1in).

Stainless steel

Stainless-steel tubing is not as common as copper, but is available in the same sizes. You may have to order it from a plumbers' merchant. It's harder than copper, so cannot be bent as easily, and is difficult to solder. It pays to use compression joints to connect stainless-steel pipes, but tighten them slightly more than you would when joining copper.

Stainless steel does not react with galvanized steel (iron) – see Electro-chemical action (bottom left).

Lead

Lead is never used nowadays for any form of new plumbing – but there are thousands of houses that still have a lead rising main connected to a modernized system.

Lead plumbing that's still in use will be nearing the end of its life, so replace it as soon as an opportunity arises. When drinking water lies in a lead pipe for some time, it absorbs toxins from the metal. If you have a lead pipe supplying your drinking water, always run off a little water before you use any.

Galvanized steel (iron)

Galvanized steel was once commonly used for supply pipes, both below and above ground, having taken over from lead. It was then superseded by copper.

There are two problems with this type of pipe. It rusts from the inside and resists water flow as it deteriorates. Also, when it is joined to copper, the galvanizing breaks down rapidly because of an electrolytic action between the copper and zinc coating.

Cast-iron waste pipes
All old soil pipes are made from cast iron, which is prone to rusting. If it weren't for their relatively thick walls, pipes of this kind would have rusted away long ago. Should you need to replace a cast-iron pipe, ask for one of the plastic alternatives.

Copper pipes
The economic choice for modern plumbing systems.

Stainless steel
Owing to its superior appearance and strength, stainless steel is used where pipe runs are exposed. It does not cause electrolytic action with galvanized-steel pipes.

Lead
This is still found in older houses. It can introduce toxins into the drinking-water supply, so should be replaced.

Iron
Iron pipes are used for mains water supply in some older systems. Iron is susceptible to furring-up and decay, which can result in low water pressure and leaks. Cast iron is used for waste pipes in older buildings.

SEE ALSO > Soldered joints 21, Push-fit joints 22, Plastic waste pipes 24, Soft water 50, Supplementary bonding 72, Main switch equipment 76

Metal joints and fittings

Joints are made to connect pipes at different angles and in various combinations. There are adaptors for joining metric and imperial pipes, and for connecting one kind of material to another. You need to consult manufacturers' catalogues to see every variation, but the examples on this page illustrate a typical range of joints. Plumbing fittings such as valves are made with demountable compression joints, so that they can be removed easily for servicing or replacement.

Straight connectors
To join two pipes end to end in a straight line.
1 For pipes of equal diameter – compression joint.
2 Reducer to connect a 22mm (³⁄₄ in) pipe to a 15mm (¹⁄₂ in) pipe – capillary joint.

Bends or elbows
To join two pipes at an angle.
3 90° – compression joint.

Tees (T-joints)
To join three pipes.
4 Equal tee, for joining three pipes of the same diameter – capillary joint.
5 Unequal tee, for reducing size of pipe run when connecting a branch pipe – compression joint.

Adaptors
To join dissimilar pipes.
6 Straight coupling for joining 22mm and ³⁄₄ in pipes – compression joint.
7 Connector for joining copper to galvanized steel – compression joint for copper, threaded female coupling for steel.

Fittings
Identical jointing systems are used to connect fittings.
8 End cap, to seal pipes – compression joint.
9 Tap connector for connecting supply pipe to tap – capillary joint.
10 Tank connector, joins pipes to cisterns – compression joint.
11 Bib-tap wall plate, for fixing tap on outside wall.
12 Bib tap has threaded tail to fit wall plate.
13 Gate valve to fit in straight pipe run – compression joint.
14 Draincock for emptying a pipe run – compression joint.
15 Straight service valve for isolating a tap or float valve – compression joint.
16 Double-check non-return valve, used for outside taps and other outlets where contamination of water supply is possible – compression joint.

1 Equal-size connector

2 Reducer

3 Elbow 90°

4 Equal tee

5 Unequal tee

6 Straight coupling

7 Copper-to-steel connector

8 End cap

9 Tap connector

10 Tank connector

11 Bib-tap wall plate

12 Bib tap

13 Gate valve

14 Draincock

15 Straight service valve

16 Double-check non-return valve

Pipe joints

It's not possible to make a strong, watertight joint by simply soldering pipes together, so capillary or compression joints are used to connect pipes or attach fittings such as tap connectors, valves and the like.

Capillary joints

Capillary joints are made to fit snugly over the ends of a pipe. The very small gap between the pipe and joint sleeve is filled with molten solder - when this cools, it holds the joint together and makes it watertight. Capillary joints are neat and inexpensive – but because you need to heat the metal with a gas torch, there is a risk of fire when working in confined spaces under floors.

Soldering capillary joints

Solder is introduced to each mouth of the assembled end-feed joint and flows by capillary action into the fitting.

The rings pressed into the sleeves of an integral-ring fitting contain the exact amount of solder to make perfect joints.

Compression joints

Compression joints are simple to use, but more expensive than capillary joints. They are also more obtrusive, and you may find it hard to manoeuvre a wrench where space is restricted. When the cap-nut is tightened with a wrench it compresses a ring of soft metal, known as an olive, to fill the joint between fitting and pipe.

Corrosion resistance
Corrosion can take place between brass fittings and copper pipes. Look for the symbol that denotes corrosion-resistant brass fittings.

CAP-NUT
COUPLING BODY
CUT PIPE SQUARE PRIOR TO ASSEMBLY
OLIVE
OLIVE

SEE ALSO > Metal plumbing 19

Cutting metal pipe

Calculate the length of pipe you need, allowing enough to fit into the sleeve of the joint at each end. Whatever type of joint you use, it's essential to cut the end of every length of pipe square.

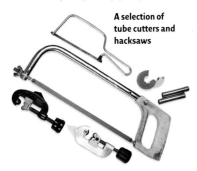

A selection of tube cutters and hacksaws

To ensure a perfectly square cut each time, use a tube cutter. Align the cutting wheel with your mark, and adjust the handle of the tool to clamp the rollers against the pipe (**1**). Rotate the tool around the pipe, adjusting the handle after each revolution to make the cutter bite deeper into the metal.

A tube cutter makes a clean cut on the outside of the pipe, but use the pointed reamer on the tool to clean the burr from inside the cut end (**2**).

If you use a hacksaw, make sure the cut is square by wrapping a piece of paper with a straight edge around the pipe. Align the wrapped edge and use it to guide the saw blade (**3**). Remove the burr, inside and out, with a file.

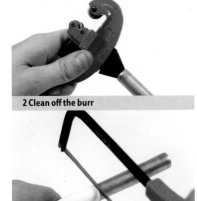

1 Clamp the tube cutter onto the pipe

2 Clean off the burr

3 Wrap paper around the pipe to guide a saw

Making soldered joints

Soldering pipe joints is easy once you have had a little practice. The fittings are cheap, so you can afford to try out the techniques before you begin to install pipework. You need a gas torch to apply heat, some flux to clean the metal, and solder to make the joint. Make sure the pipe is perfectly dry before you attempt to solder a joint.

Solder and flux

Solder is a soft alloy manufactured with a melting point lower than that of the metal it is joining. Plumbers' solder is sold as wound wire.

Copper must be spotlessly clean and grease-free if it is to produce a properly soldered joint. Even when cleaned mechanically with wire wool, copper begins to oxidize immediately; a chemical cleaner known as flux is therefore painted onto the metal to provide a barrier against oxidation until the solder is applied. A non-corrosive flux in the form of a paste is the best one to use. On stainless steel use a highly efficient active flux – but wash it off with warm water after the joint is made, or the metal will corrode.

Gas torches

To heat the metal sufficiently for a soldered joint, most plumbers use a gas torch. Gas, liquefied under pressure, is contained in a disposable metal canister. When the control valve of the torch is opened, gas is vaporized to combine with air, making a highly combustible mixture. Once ignited, the flame is adjusted until it burns steadily with a clear blue colour.

Using end-feed joints

Clean the pipe with wire wool (**1**) then apply flux (**2**). Assemble the components (**3**), then heat the joint evenly. When the flux begins to bubble, remove the flame and touch the solder to two or three points around the fitting (**4**) – the joint is full of solder when a bright ring appears around each sleeve. Allow it to cool.

1 Clean the pipe with wire wool

3 Push components together

2 Apply flux

4 Heat and apply solder

Using integral-ring joints

Clean the ends of each pipe and the inside of the joint sleeves with wire wool or abrasive paper until the metal is shiny. Brush flux onto the cleaned metal and push the pipes into the joint, twisting them to spread the flux evenly. Push each pipe up against the stop in the joint.

If using elbows or tees, mark the pipe and joint with a pencil, to make sure they do not get misaligned during soldering.

Slip a ceramic tile or a plumber's fibre-glass mat behind the joint to protect flammable materials, then apply the flame of a gas torch to the area of the joint to heat it evenly. When a bright ring of solder appears at each end of the joint, remove the

flame and allow the metal to cool for a couple of minutes before disturbing it.

Repairing a weeping joint
When you fill a new installation with water for the first time, check every joint to make sure they're watertight. If you notice water 'weeping' from a soldered joint, drain the pipe and allow it to dry. Heat the joint and apply some fresh solder to the edge of each mouth. If it leaks a second time, heat the joint until you can pull it apart with gloved hands. Either use a new joint or clean and flux all surfaces and reuse the same joint, adding solder as if you were working with a new end-feed fitting.

SEE ALSO > Pipe fittings 20

Compression joints

Using compression fittings is so straightforward that you will be able to make watertight joints without any previous experience. All that's required for success is to ensure that the pipes to be joined are clean and that the fitting is not overtightened.

Assembling a joint

Straight connector
Compression joint to join two pipes of equal diameter, end to end, in a straight line.

Elbow joint
A 90° elbow compression joint connects two pipes at an angle.

Cut the ends of each pipe square and clean them, along with the olives, using wire wool. Dismantle a new joint and slip a cap-nut over the end of one pipe, followed by an olive (**1**). Look carefully to see if the sloping sides of the olive are equal in length. If one is longer than the other, that side should face away from the nut.

Push the pipe firmly into the joint body (**2**), twisting it slightly to ensure it is firmly against the integral stop. Slide the olive up against the joint body, then tighten the nut by hand.

The olive must be compressed by just the right amount to ensure a watertight joint. As a guide, make a pencil mark on one face of the nut and on the opposing face on the joint body (**3**); then, holding the joint body steady with a spanner, use another spanner to turn the nut one complete revolution (**4**). Assemble the other half of the joint in exactly the same manner.

Some plumbers like to wrap a single turn of PTFE tape over the olive before tightening the nut, to make absolutely sure the joint is watertight. However, a properly tightened compression joint should be watertight without it.

1 Slip an olive onto the pipe after the cap-nut

3 Mark the nut and joint with a pencil

2 Clamp the joint to the pipe with the nut

4 Tighten the joint with two spanners

Fixing weeping joints

Crushing an olive by overtightening a compression joint will cause it to leak. Drain the pipe and dismantle the joint. Cut through the damaged olive with a junior hacksaw, taking care not to damage the pipe. Remake the joint with a new olive, restore the supply of water, and check for leaks once more.

Saw through a damaged olive

Metal push-fit joints

Metal push-fit joints are available as an alternative to the plastic variety. Made from brass alloy, they have a one-piece body that incorporates a stainless-steel grab ring to prevent the fitting coming loose under pressure. A release mechanism is incorporated to make them easy to remove. They are quick to use, less bulky than their plastic counterparts and better able to withstand higher water pressures, but they are more expensive, especially if you are carrying out a major project.

SEE ALSO > Metal joints and fittings 20, Plumbing adaptors 24

Bending pipes

You can change the direction of a pipe run by using an elbow joint, but there are occasions when bending the pipe itself will produce a neater or more accurate result. A simple bending spring is all that is required to make bends in small copper pipework; you can hire or buy special tools for more substantial stock.

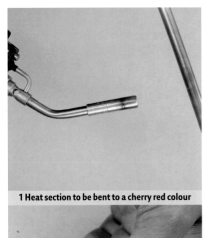

1 Heat section to be bent to a cherry red colour

Plumbers' bending springs

If you want to carry a pipe over a small obstruction (another pipe, for example), a slight kink in the pipe will be less of an obstruction to the flow of water and create less noise than two elbows within a few centimetres of each other. If you need to to run pipes into a window alcove where the walls meet at an unusual angle, bending the pipes accurately will allow you to fit the pipes neatly against the alcove walls.

Using a bending spring

A bending spring is the cheapest and easiest tool for making bends in small pipe runs. It is a hardened-steel coil spring that supports the walls of copper tube to stop it kinking. Most bending springs are made to fit inside the pipe, but some slide over it.

Soften the section to be bent by heating first (**1**) and cooling, then slide the spring (see below) into the tube (**2**), so it supports the area to be bent. Hold the tube against your padded knee and bend to required angle (**3**). The tube will grip the spring, but slipping a screwdriver into the ring at one end and turning it anticlockwise will compress the spring so it can be pulled out.

If you make a bend some distance from the end of a tube, you won't be able to withdraw the bending spring with a screwdriver. Either use an external spring or tie string to the ring and lightly grease the spring with petroleum jelly before you insert it. Slightly overbend the tube, open it out to the correct angle to release the spring, then pull it out with the string.

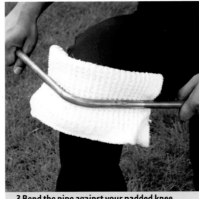

2 Lubricate spring and tie string to end

3 Bend the pipe against your padded knee

Using a pipe bender

Although you can hire bending springs to fit the larger pipes, it isn't easy to bend 22 or 28mm (¾ or 1in) tube over your knee – so it is well worth hiring a pipe bender to do the job. Ensure that you fit the correct former to suit the pipe diameter.

Hold the pipe against the radiused former and insert the straight former to support it. Pull the levers towards each other to make the bend, and then open up the bender to remove the pipe.

Getting the bends in the right place

It is difficult to position two or more bends accurately along a single length of pipe. In an alcove, it's easier to bend lengths of pipe to fit each corner, then cut the tubes where they overlap and insert joints.

1 A pipe bender makes accurate bends

2 Pull the handles together to bend the pipe

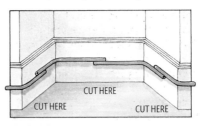

CUT HERE
CUT HERE   CUT HERE

Formers are marked to aid bends to common angles

SEE ALSO > Connecting plastic to metal plumbing 25

Plastic plumbing

Plastic plumbing is lightweight and extremely simple to assemble. It doesn't burst when frozen, corrode, or adversely affect other materials; and, depending on the type of plastic, it can be used both for cold water and hot, including central-heating pipework. Most plastic systems can be connected to existing metal pipes.

Plastic joints and fittings are similar to the ones used for metal plumbing, but are typically larger in size. Joints and pipes are for the most part manufactured from the same material, but there are several specialized connectors available for joining plastic plumbing to taps, tanks and existing metal plumbing. To see the huge variety of plastic joints, you need to browse through manufacturers' catalogues, but the selection below shows some of the most common types.

Straight connectors
For joining two pipes end to end.
1 For pipes of equal diameter – push-fit.

Elbows
For joining two pipes at an angle.
2 Elbow 45° – solvent weld.
3 Elbow 90° – push-fit.

Adaptors
To join dissimilar pipes.
4 Plastic-to-copper connector – push-fit and compression joint.

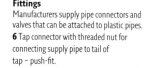

Tees
For joining three pipes.
5 Equal tee for joining 15mm (½ in) branch pipes – push-fit.

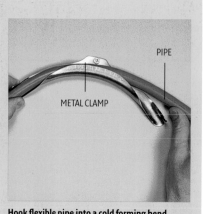

Fittings
Manufacturers supply pipe connectors and valves that can be attached to plastic pipes.
6 Tap connector with threaded nut for connecting supply pipe to tail of tap – push-fit.
7 Tank connector joins pipes to storage tanks and cisterns – push-fit.
8 Stopcock – push-fit.

Plastic supply pipes

Plastic supply pipes are made to the same standard sizes as metal pipework, but there may be a slight variation in wall thickness from one manufacturer's stock to another.

Chlorinated polyvinyl chloride (cPVC)
A versatile plastic suitable for hot and cold supply. It can even withstand the temperatures that are required for central-heating systems.

Polybutylene (PB)
A tough, flexible plastic pipe used for hot and cold supply, and central heating. Available in standard lengths or continuous coils, PB resists bursting when frozen. It will sag if unsupported.

Cross-linked polyethylene (PEX)
Although it expands considerably when it is heated, PEX is used to make pipes that supply hot and cold water and for under-floor heating systems. However, it tends to sag, so is unsuitable for surface running. A PEX pipe resists bursting when subjected to frost. Twin-wall PEX, with an oxygen-diffusion barrier (see far left) in the form of an aluminium layer sandwiched between the walls, is semi-rigid.

Medium-density polyethylene (MDPE)
Widely used for underground domestic supply pipes, the pipes, normally coloured blue, can be laid in continuous lengths and are resistant to pressure and corrosion.

• Oxygen-diffusion barriers
There's some concern that a small amount of oxygen drawn through the walls of plastic central-heating pipes contributes to the corrosion of the system. To prevent this happening, an oxygen-diffusion barrier is built into the walls of the pipe.

Bending plastic pipes

Flexible pipes can be bent cold to a minimum radius of eight times the pipe diameter. Use a pipe clip at each side of the bend to hold the curve, or use a plastic or metal cold forming bend. It is easy to thread flexible pipe around obstacles or under floorboards.

Rigid plastic pipe can be bent by heating it gently with a gas torch. Keep the flame moving and revolve the pipe. When the pipe is soft enough, bend it by hand on a flat surface. Hold it still till the plastic hardens again. Wear thick gloves when handling hot plastic.

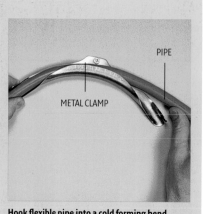

PIPE

METAL CLAMP

Hook flexible pipe into a cold forming bend

cPVC

PB

PEX

MDPE

SEE ALSO > Solvent-weld joints 27

Joining plastic supply pipes

Some plastic supply pipes can be connected using solvent-weld joints (as described for waste systems), but it is easier and more convenient to use the push-fit connectors shown below.

Push-fit joints

When the pipe is inserted, an O-ring seals in the water in the normal way and (depending on the model) a special plastic grab ring, or a collet with stainless-steel teeth, grips the tube securely to prevent water under mains pressure forcing the joint apart. Joints fitted with collets can be disconnected easily, but to dismantle the other type of push-fit joint, it's necessary to remove the retaining cap and prise open the grab ring, using a special tool.

Push-fit joints are more obtrusive than their solvent-welded equivalents – but are much faster to assemble.

Grab-ring push-fit joint
A grab ring holds the pipe, to resist water under pressure.

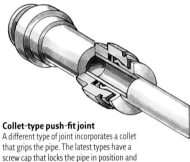

Collet-type push-fit joint
A different type of joint incorporates a collet that grips the pipe. The latest types have a screw cap that locks the pipe in position and applies more compression to the rubber seal.

Connecting metal to plastic plumbing

Special adaptor couplings are needed in order to connect most types of plastic pipe to copper or galvanized-steel plumbing. To join polybutylene pipe to copper, insert a metal support sleeve, then use a standard brass compression joint; or use a push-fit connector to join copper pipes to a polybutylene run. Cut and deburr the copper pipe carefully before pushing it into the joint.

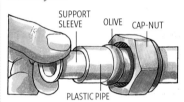

SUPPORT SLEEVE · OLIVE · CAP-NUT · PLASTIC PIPE

Joining plastic pipe with a compression fitting
Insert support sleeve before tightening the joint.

• **Repairing a weeping joint**
A push-fit joint on a supply pipe may leak if the pipe is not pushed home fully, or if the O-ring is damaged. A plastic pipe that has not been cut square or has scratches in the joint area may leak, too. Take the joint apart and check all the components, replacing any that are damaged, then reassemble carefully, pushing the pipe fully home.

Using grab-ring joints

Cut polybutylene pipe to length with the special shears that are supplied by the manufacturer (**1**); alternatively, use a sharp craft knife. Provided that you make the cut reasonably square, the joint will be watertight.

Push a metal support sleeve into the pipe (**2**), and, if necessary, smear a little silicone lubricant around the end of the pipe and inside the socket (**3**).

Push the prepared pipe firmly a full 25mm (1in) into the socket (**4**). As the joint can revolve freely around the pipe after connection without breaking the seal, there is no problem when aligning tees and elbows with other pipe runs.

1 Cut pipe to length	**2 Insert metal sleeve**	**3 Apply lubricant**	**4 Push pipe into joint**

Dismantling a joint
To dismantle a joint, unscrew the cap and pull out the pipe. Slide off the rubber O-ring, then prise off the grab ring, using

Prise open the grab ring, using a special tool

a special demounting tool (see below left). Grab rings are not reusable and you must use new ones when re-assembling the joint using the technique described above. Never try to assemble the fitting like a compression joint, or it will blow out under pressure.

If the type of fitting is no longer available, use another brand that closely matches the original. Given that there are small differences between brands of pipe, there is a risk that the new fitting won't seal properly, so you may have to replace the length of pipe and several fittings.

Collet-type joints

Push-fit joints that incorporate collets are particularly easy to assemble. Cut the end of the pipe square, push it into the socket until it comes up against the internal stop, then pull back to check that it is secure.

If you need to dismantle a joint, hold the collet in with your fingertips (**1**) and pull the pipe out of the socket.

Join metal pipes the same way, but remove burrs and sharp edges to prevent tearing the O-ring. Provide extra grip by twisting the outer collar to lock it (**2**).

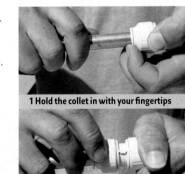

1 Hold the collet in with your fingertips

2 Twist collar to lock joint

Cutting plastic pipe
Polybutylene pipe is easy to cut, using special shears.

SEE ALSO > Adaptor couplings 24, Solvent-weld joints 27, Supporting pipes 28

Plastic waste pipes

Plastics are complex materials, each having its own properties. Consequently, a technique or material that is suitable for joining one plastic may not be suitable for another. To make watertight joints, it's vital to follow the manufacturer's instructions carefully, and to use the recommended solvents and lubricants.

Types of plastic

Plumbing manufacturers have a wide variety of plastics to draw upon, each with its own special characteristics.

Modified unplasticized polyvinyl chloride (MuPVC)

A hard plastic, used for solvent-weld waste pipe and fittings. It is resistant to most domestic chemicals, and is not affected by ultra-violet light when used outdoors. It is slightly more flexible than uPVC, which is used for soil pipes with push-fit and solvent-weld joints.

Polypropylene (PP)

A slightly flexible plastic with a slightly waxy feel, used for waste systems. It's impossible to glue PP, so it is assembled with push-fit joints.

Acrylonitrile butadiene styrene (ABS)

A very tough plastic that is equally suited to hot and cold waste. It can be either solvent-welded or compression-jointed.

Joints and fittings

As well as the usual types of joint, waste systems also include easy-flow swept bends and tees for efficient drainage. The swept part of the fitting should face downhill to aid water flow.

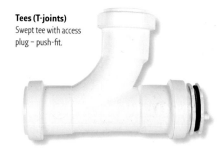

Tees (T-joints)
Swept tee with access plug – push-fit.

Fittings
Bottle trap for sink or basin – compression joint.

Branch 45°–
solvent-weld.

Bends and elbows
Elbow 90°–
push-fit.

Bend 90°–
solvent-weld.

PLASTIC WASTE-PIPE SIZES

Overflow pipes	22mm (3/4 in)
Washbasin waste pipes	32mm (1 1/4 in)
Bath/shower and sink waste pipes	40mm (1 1/2 in)
Soil pipe	100mm (4in)

Waste-pipe joints

Solvent-weld joints

Lengths of pipe are linked by simple socketed connectors coated with solvent which dissolves the surfaces of the mating components. As the solvent evaporates, the joints and pipes are literally fused together into one piece of plastic.

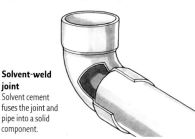

Solvent-weld joint
Solvent cement fuses the joint and pipe into a solid component.

Compression joints

So that they can be dismantled easily, sink, bath and washbasin traps are often connected to the pipework by means of compression joints that use a rubber ring or washer to make the joint watertight.

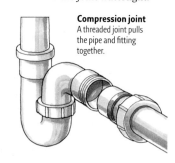

Compression joint
A threaded joint pulls the pipe and fitting together.

Push-fit joints

Because a waste system is never under pressure, a pipe run can be constructed by simply pushing plain pipes into the sockets of the joints. A captive rubber seal in each socket holds the pipe in place.

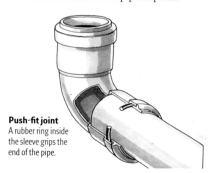

Push-fit joint
A rubber ring inside the sleeve grips the end of the pipe.

SEE ALSO > Drainage systems 15, Plastic plumbing 24, Joining plastic supply pipes 25

Joining plastic pipes

It's important to follow the instructions supplied with any particular brand of pipe or fitting, but the methods given below and on the facing page describe the basic techniques for connecting plastic pipes. Work carefully and avoid spilling solvent cement – it will etch the surface of the pipework and damage some other plastics, as well.

Making solvent-weld joints

While the sequence of illustrations on the right shows large-diameter waste pipe, the methods described are equally valid for joining plastic supply pipe.

Cut the pipe to length with a fine-tooth saw, allowing for the depth of the joint socket. To make sure your cut is square, wind a piece of notepaper round the tube, aligning the wrapped edge as a guide (**1**). Revolve the pipe away from you as you cut it. Smooth the end with a file (**2**).

Welding the joint

Push the pipe into the socket to test the fit, then mark the end of the joint on the pipe with a pencil (**3**). This will act as a guide for applying the solvent. You need to key both the outside of the pipe and the inside of the socket with fine abrasive paper before using some solvents (check the manufacturer's instructions).

Before dismantling elbows and tees, mark the pipe and joint with a pencil (**4**), to help you align them correctly when you reassemble the components.

Use a clean rag to wipe the surface of the pipe and fitting with the recommended spirit cleaner. Brush solvent evenly onto both components (**5**), then immediately push home the socket. (Some manufacturers recommend that you twist the joint to spread the solvent.) Align the joint properly and leave it for 15 seconds.

The pipe is ready for use with cold water after an hour. But don't pass hot water through the system until at least four hours have elapsed (depending on the manufacturer's recommendations) or, preferably, longer.

Repairing a weeping joint

If a joint leaks, leave it to dry out naturally. Then apply a little more of the solvent cement to the mouth of the socket, allowing it to flow into the joint by capillary action.

You will have to drain a supply pipe before you can make this repair.

1 Use paper as a guide to keep the cut square

2 Smooth the end with a file

3 Assemble the joint and mark the pipe

4 Mark the joint so it can be realigned

5 Paint solvent up to the pencil mark

Making push-fit joints

Cut the pipe to length and chamfer the end, as for solvent-weld joints. Wipe the inside of the socket with the recommended cleaner, and lubricate the pipe with some silicone lubricant.

Push the pipe into the joint right up to the stop, and mark the edge of the socket on the pipe with a pencil (**1**).

Withdraw the pipe about 9mm (³⁄₈in) (**2**), to allow the pipe to expand when subjected to hot water.

1 Mark the edge of the socket on the pipe

2 Withdraw the pipe about 9mm (³⁄₈in)

Repairing a weeping push-fit joint

A push-fit joint will leak if the rubber seal has been pushed out of position. Dismantle the joint and check the condition of the seal.

Making compression joints to traps

Traps with compression joints are made for connecting directly to a plain waste pipe (see opposite). Just slip the threaded nut onto the waste pipe, followed by the washer and then the rubber ring. Push the pipe into the socket of the trap and tighten the compression nut.

SEE ALSO > Repairs and maintenance 9, Plastic plumbing 24, Joining plastic supply pipes 24

Running and concealing pipework

The most difficult part of many plumbing jobs is not the actual plumbing itself but the work required to run and conceal the associated pipework. This may involve lifting floorboards and drilling through walls and joists, so spend time working out which is the least disruptive route. You may also have to consider how to hide pipework from view where it has to be run across a wall.

● **Supporting pipe runs**
The spacing of pipe clips will vary depending on the direction the pipe is running, the material it's made from and its diameter. Horizontal runs will need more clips than vertical ones.

On a horizontal run, plastic pipes up to 15mm (½in) diameter need to be supported every 300mm (1ft); 22mm (¾in) pipes every 500mm (1ft 8in). Vertical runs require spacings of 500mm (1ft 8in) and 800mm (2ft 8in) respectively.

Horizontal runs of 15mm (½in) copper pipe should be supported every 1.5m (5ft); for 22mm (¾in) the spacing is 2m (6½ft). Vertical runs require spacings of 2m (6½ft) and 2.5m (8ft 3in) respectively.

Supporting pipework

It's important that all pipework used in waste, supply or central heating applications is supported or secured at regular intervals.

Unsecured pipework will place undue strain on joints, increasing the risk of joint failure. It may also be noisy, due to excess movement caused by expansion and contraction, and, if visible, will sag and look unsightly.

There's a wide variety of clips for securing surface run pipework. Copper pipes are usually secured with push-in plastic clips, which are screwed in place. The most common clip has an open top (see below) but some have a hinged latch which clips over the pipe for added security. Nail-in clips that resemble cable clips are available for small-bore pipes, while copper clips - known as saddles - with two screw fixings are often used on exterior runs, where their lower profile and extra strength are desirable.

Saddle-style clips are generally used for plastic waste pipe, though standard pipe clips can be used for 15mm or 22mm (½in or ¾in) plastic supply pipes.

Pipes in floors

Pipes running across the direction of the joists can be set into notches cut into the top of the timbers.

The notches should only be made in the top of the joist and should be no deeper than 0.125 times the depth of the joist to avoid weakening it. Notches should also be no closer to the joist supports than 0.07 times the span of the joist, or further than a quarter of the length of the joist away. This is to avoid the areas of maximum stress or bending in the timber.

The notches should be just wider and deeper than the diameter of the pipe to avoid contact between pipe, joist and floorboards. Lining the notch and covering the top of the pipe with foam mat will prevent noise as the pipe expands or the board above is trodden on.

If you can arrange for the pipe to be under the centre line of the board covering it, you can nail the board back either side of the notch.

Drilling joists

If you have to drill holes through the joist for pipes - a convenient way to run flexible plastic pipes - then the holes should be drilled on the centre line of the joist and have a diameter no bigger than a quarter of the joist's depth. Multiple holes should be spaced at least three times the largest diameter permitted apart.

Drilling large holes between joists can be tricky as there isn't enough room for the power drill and bit. You can hire or buy a special right-angled drill specially designed for this job, which will make drilling straight, level holes simple.

Pipes running parallel to joists can be clipped to the joists or supported with a 'bridge' made from battens and plywood to bridge the gap between two joists.

Once you've run a pipe under floorboards, mark its route with a felt pen or paint on the surface of the boards so you know where it is in future.

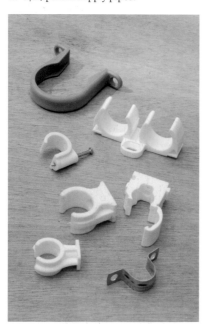

A selection of pipe clips

Chisel out the notch

Drill holes between joists with an angled drill

Pipes should lie beneath the board level

Pipes running parallel to joists can be clipped

SEE ALSO > Metal pipes 19, Plastic plumbing 24

Concealing pipework

With carefully designed pipe runs, it should be possible to plumb your house without a single pipe being visible. In practice, however, there are always situations where you have no option but to surface-run some pipes.

You can minimize the effect by taking care to group pipes together neatly and keeping runs both straight and parallel. When painted to match the skirtings or walls, such pipes are barely visible.

Alternatively, using softwood battens and plywood, you can make your own accessible ducting to bridge the corner of a room; or construct a false skirting that is deep enough to contain the pipes.

For total accessibility, you can use proprietary ducting made from PVC. This is manufactured in a range of sizes, to contain grouped or individual pipes (see below).

Make sure that gate valves or service valves remain accessible - you may have to construct removable panels in order to ensure access. These could be hinged or attach with magnetic cupboard clips.

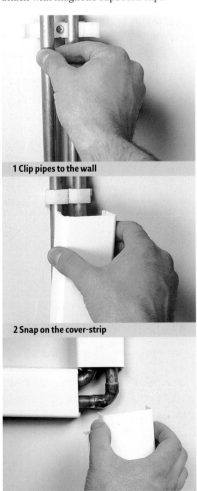

1 Clip pipes to the wall

2 Snap on the cover-strip

3 Snap on corner covers

Drilling through walls

Use a core bit for large diameter holes

Drill several small holes, then chop out waste

Where pipes have to pass through walls - perhaps for a waste from a washing machine or a supply pipe for a garden tap - you'll need a suitable hole.

For smaller pipes up to 15mm (½in) or so, you can use a large-diameter masonry bit in an ordinary power drill - the bit will have a reduced diameter shank to fit the drill's chuck.

However, larger diameter holes are best made with a core bit. These are available in a range of diameters and can even be used to cut holes for 100mm (4in) diameter waste pipe. Diamond or carbide tipped, core bits are expensive and best hired; for bigger holes you'll need to hire a suitably powerful drill as your own

probably won't be up to the job. Holes should slope slightly downward from the inside to the outside to prevent any risk of rainwater running along the pipe.

Copper pipes should be sleeved with a piece of plastic pipe for protection where they pass through the wall. Any space around the pipe can be filled with mortar, silicone sealant or expanding foam filler.

If you don't want to hire a core drill, you can form a larger hole by drilling a series of small holes with a standard masonry bit, then using a hammer and cold chisel to chop out the masonry between the holes. Cut from the outside in to avoid any risk of breaking the surface brickwork around the hole.

Waste pipe to soil pipe

A proprietary pipe boss is used to connect a basin waste pipe to a single-stack plastic soil pipe. There are various ways of connecting the boss, one of the simplest being to clamp it with a strap.

Mark where the basin waste meets the soil pipe, and use a hole saw to cut a hole of the recommended diameter (**1**). Smooth the edge of the hole with abrasive paper.

Wipe both contacting surfaces with the manufacturer's cleaner, then apply gap-filling solvent cement around the hole. Strap the boss over the hole and tighten the bolt (**2**).

Insert the rubber lining in the boss, in preparation for the waste pipe (**3**).

Lubricate the end of the pipe and push it firmly into the boss (**4**). Clip the pipe to the wall.

1 Cut a hole in the pipe with a hole saw

3 Insert the rubber lining

2 Strap the boss over the hole

4 Push the waste pipe into the boss

Allow space for insulation
Exterior supply pipes must be insulated and standard pipe clips don't leave enough room between the wall and the pipe for foam insulation. You can buy spacers to suit most pipe clips which distance the clip from the wall and leave space for the insulation to fit comfortably.

SEE ALSO > Cleansing the waste pipe 16, Metal pipes 19, Bending pipes 23, Plastic plumbing 24–5

Selecting taps

Taps can be fashionable as well as functional and come in a huge variety of different styles and finishes. Not all taps are built to last, so check the quality if you are buying for the long term. Chromium-plated brass taps are the most durable. Check that the taps you are considering will fit the holes in your chosen appliance.

Single-lever mixer tap
Moving the lever up and down turns the water on and off. Swinging it from one side to the other alters the temperature, by mixing the hot water with the cold. These taps have smaller, 10mm (3/8in) feed pipes, which will need adaptors to connect to 15mm (1/2in) supply pipes.

● The right pressure
Some taps imported from the Continent have relatively small inlets and are intended for use with mains-pressure supply only. These taps will not work efficiently if they are connected to a low-pressure tank-fed supply.

Basin and bath taps
(top row – left to right)
One-hole basin mixer
Single shrouded head pillar taps
Single-lever mixer with pop-up waste

(bottom row – left to right)
Shower-mixer deck
Monobloc bath mixer
Remote spout bath mixer

Types of tap

The majority of washbasins are fitted with individual taps for hot and cold water. While capstan-head taps are still manufactured for use in period-style bathrooms, most modern taps have a shrouded head made of metal or plastic.

A lever-head tap turns the water from off to full on with one quarter turn only, useful for those who may have difficulty in manipulating other taps.

In a mixer tap, hot and cold water are directed to a common spout. Water is supplied at the desired temperature by adjustment of the two valves. With a single-lever mixer tap, flow rate and temperature are controlled by adjusting one lever, which moves up to control the flow and sideways to alter temperature.

Washbasin mixer taps sometimes incorporate a pop-up waste plug. A series of interlinked rods, operated by a button or small knob on the centre of the mixer, open and close the waste plug in the basin.

Normally, the body of the tap (which connects the valves and spout) rests on the upper surface of the washbasin. But it is also possible to mount it in its entirety on the wall above the basin. Another alternative is for the valves to be mounted on the basin and divert hot and cold water to a spout mounted on the wall above.

Tap mechanisms

Over recent years there have been some revolutionary changes in the design of taps, which have made them easier to operate and simpler to maintain.

Rising-spindle taps
This traditional tap design has a washer on the end of a spindle that rises as the tap is turned on. It is a simple, rugged mechanism that lasts for years.

Non-rising-spindle taps
Theoretically, these taps should exhibit fewer problems than rising-spindle taps, because the mechanism imposes less wear on the washer. In practice, however, the spindle's fine thread is prone to wear, and misalignment may be caused by the circlip that holds the mechanism in place.

Ceramic-disc taps
With these taps, precision-ground ceramic discs are used in place of the traditional rubber washer. One disc is fixed and the other rotates until the waterways through them align and water flows. There is minimal wear, as hard-water scale or other debris is unlikely to interfere with the close fit of the discs. However, if a problem does develop, the entire inner cartridge and the lower seal can be replaced.

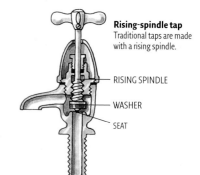

Rising-spindle tap
Traditional taps are made with a rising spindle.

RISING SPINDLE
WASHER
SEAT

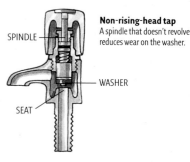

Non-rising-head tap
A spindle that doesn't revolve reduces wear on the washer.

SPINDLE
WASHER
SEAT

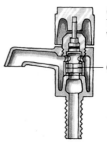

Ceramic-disc tap
The rubber washer is replaced with rotating ceramic discs.

CERAMIC DISCS

SEE ALSO > Repairing taps 10–11

Removing old taps

When replacing taps, you will want to use the existing plumbing if possible, but disconnecting old, corroded fittings can be difficult.

Apply some penetrating oil to the tap connectors and to the back-nuts that clamp the tap to the basin. While the oil takes effect, shut off the cold and hot water supply to the taps.

If necessary, apply heat with a gas torch to break down the corrosion – but wrap a wet cloth around nearby soldered joints, or you may melt the solder. Take care that you do not damage plastic fittings and pipes, and protect flammable surfaces with a ceramic tile. Try not to play the flame onto a ceramic basin.

Cranked spanners

It is not always possible to engage the nuts with a standard wrench. Instead, use a special cranked spanner designed to reach into the confined spaces below a basin or bath. You can apply extra leverage to the spanner by slipping a stout metal bar or wrench handle into the other end.

A cranked spanner fits basin and bath taps

Removing a stuck tap

Even when you have disconnected the pipework and back-nut, you may find that the taps are stuck in place with putty. Break the seal by striking the tap tails lightly with a wooden mallet. Clean the remnants of putty from around the holes in the basin, then fit new taps. If the tap tails are shorter than the originals, buy special adaptors designed to take up the gaps.

Releasing a tap connector
Use a special cranked spanner to release the fixing nut of a tap connector.

Fitting new taps

If you're installing a new bath or basin, it's much easier to fit the taps before the sanitaryware is in position since access to hard-to-reach nuts will be much better.

Most taps are made to standard sizes, so you shouldn't have any difficulty with new baths or basins, but if you're replacing the taps on an older fitting, check that they will cover the holes adequately or, if you're using a two-hole tap, fit in the holes. Check that the spout is long enough to deliver water properly – there must be enough overhang to wash your hands underneath a basin tap, for instance.

Basin taps

Taps are supplied with flexible anti-rotation washers to stop the tap moving as it is turned on or off. Slip the washer onto the tap tail (**1**) and insert the tap through the hole (put the hot tap on the left). Add the back-nut (**2**) then hold the tap steady as you tighten it with a cranked spanner.

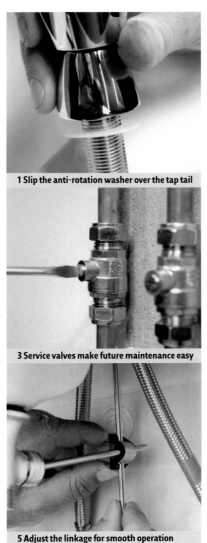

1 Slip the anti-rotation washer over the tap tail

3 Service valves make future maintenance easy

5 Adjust the linkage for smooth operation

While you have the water turned off, fit service valves so that the taps can be serviced in future without having to turn off the water. These can be fitted in the supply pipe or you can get versions that double as tap connectors (**3**).

Single-hole mixers often come with small-bore flexible hose (**4**) tails, which make connecting the supply pipes much easier.

Some mixers feature a control that operates a pop-up waste. Adjust the rods carefully to ensure (**5**) that the plug seats and lifts correctly.

Bath taps

Two-hole mixers are fitted in much the same way as basin taps. If you're fitting *in situ*, remove the overflow fitting to give you more access. Mixers are supplied with flexible washers to seal the holes – if the washer colour doesn't suit your decor, use silicone sealant instead (**6**).

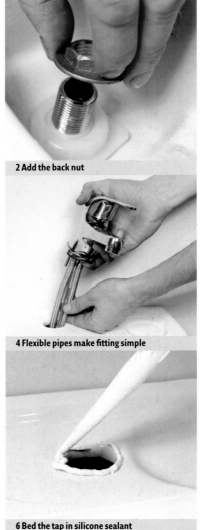

2 Add the back nut

4 Flexible pipes make fitting simple

6 Bed the tap in silicone sealant

SEE ALSO > Turning off the water 8, Connecting pipes 21–2, 25

Choosing a washbasin

Whether you're modifying existing plumbing or running pipework to a new location, fitting a washbasin in a bathroom or guest room is likely to present few difficulties provided you give some thought to how you will run the waste to the vertical stack. The pipe must have a minimum fall or slope of 6mm (¼in) for every 300mm (1ft) of pipe run and should not be more than 3m (10ft) long.

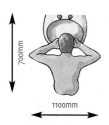

Space for a basin
Recommend dimensions for comfortable use

Selecting a washbasin

Wall-hung and pedestal washbasins are invariably made from vitreous china, but basins that are supported all round by a counter top are also available in pressed steel and plastic.

Select the taps at the same time, to ensure that the basin of your choice has holes at the required spacing to receive the taps – or no holes at all if the taps are to be wall-mounted.

Pedestal basins
The hollow pedestal provides some support for the basin and it conceals the unsightly supply and waste pipes. The basin is secured to the wall with screws, but the pedestal supports much of the weight.

Wall-hung basins
Older wall-hung basins are supported on large, screw-fixed brackets, but a modern concealed mounting is just as strong provided the wall fixings are secure. Check that you can screw into the studs of a timber-frame wall or hack off the lath-and-plaster and install a mounting board. If you want to hide pipes, consider some form of panelling.

Pedestal basin

Wall-hung basin

Corner basins
Handbasins that fit into the corner of a room are space-saving, and the pipework can be run conveniently through adjacent walls or concealed by boxing them in across the corner.

Recessed basins
In a cloakroom or WC where space is very limited, a small handbasin can be recessed into one of the walls. Also, you can recess a standard basin to conceal the plumbing.

Counter-top basins
In a large bathroom or bedroom, you can fit a washbasin or pair of basins into a counter top as part of a built-in vanity unit. Cupboards below provide storage.

Recessed basin

Counter-top basin

Corner basin

Removing a washbasin

If you want to use existing plumbing, loosen the compression nuts on the tap tails (**1**) and then remove the trap and waste pipes (**2**). Otherwise, cut through the waste and supply pipes at the point where you can most easily connect new plumbing.

Next, remove the pedestal (**3**), which may be screwed to the floor, and then undo any screws or brackets holding the basin to the wall. These may be corroded and hard to remove – there may be no option but to lever the basin off the wall.

1 Undo the nuts securing the pipes to the taps

2 Remove the trap and waste pipes

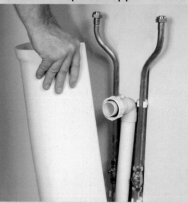

3 Lift out the pedestal after removing fixings

SEE ALSO > Draining the system 8, Selecting taps 30, Connecting a washbasin 33

Connecting a washbasin

Plan the location for your basin carefully, ensuring that there is sufficient space around it for comfortable use – ideally in an area at least 800mm (2ft 8in) wide and 700mm (2ft 4in) deep. If wall-mounted, it should be no higher than 1100mm (3ft 10in) to be suitable for all members of the family. Wall-mounted basins need to have subtantial fixings – ensure the wall is in good condition, with something solid to take the weight.

Fixing basin to the wall

Get an assistant to hold a wall-hung basin against the wall at the required height while you use a spirit level to check that it is horizontal.

Mark the fixing holes for the basin (**1**), then drill and plug the holes (**2**). Insert the special screw-in bolts and place the sink over the bolts. Add the washers and tighten the nuts (**3**).

For a pedestal basin, place the pedestal in position, then sit the basin on it and mark the fixing holes.

1 Mark the fixing holes on the wall

2 Drill the holes with a masonry bit

3 Fix the basin to the wall

Fitting trap and waste

Fit the waste outlet into the bottom of the basin as described for taps, using washers or a silicone sealant to form a watertight seal. The basin will probably have an integral overflow running to the waste – if this is the case, ensure that the slot in the waste outlet aligns with the overflow. Tighten the back-nut under the basin, while holding the outlet still by gripping its grille with long nose pliers.

If you can use the existing waste pipe, connect the trap to the waste outlet and to the end of the pipe. A two-part trap provides some adjustment for aligning with the old waste pipe.

To run a new 32mm (1¼in) waste pipe, cut a hole through the wall with a masonry core drill. Run the pipe, with sufficient fall – 6mm (¼in) per 300mm (1ft) run – to terminate over the hopper on top of the outside downpipe, or feed into a soil pipe. Fix the waste pipe to the wall with saddle clips.

Connecting the taps

You can run standard 15mm (½in) copper or plastic pipes to the taps and join them with tap connectors, but it is easier to use short lengths of flexible corrugated copper pipe designed specially for tap connection. They can be bent by hand to allow for any slight misalignment between the supply pipes and tap tails, and they are easy to fit behind a pedestal. Each pipe has a tap connector at one end and a capillary or compression joint at the other for connection to the plumbing system.

Connect the corrugated pipes to the tap tails, leaving them hand-tight only. Then run new branch pipework to meet the corrugated pipes, or connect them to the existing plumbing. Make soldered or compression joints to connect the pipes. Use a cranked spanner to tighten the tap connectors. Turn on the water supply and check the pipes for leaks; if you need to repair a weeping soldered joint, drain the system, clean the joint with wire wool and coat with flux prior to heating and adding more solder.

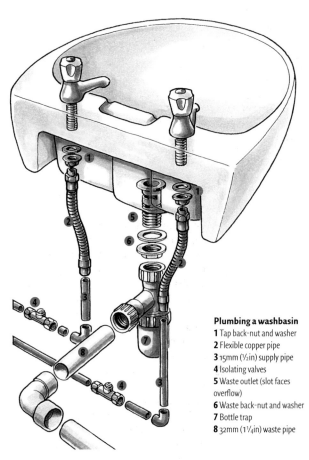

Plumbing a washbasin
1 Tap back-nut and washer
2 Flexible copper pipe
3 15mm (½in) supply pipe
4 Isolating valves
5 Waste outlet (slot faces overflow)
6 Waste back-nut and washer
7 Bottle trap
8 32mm (1¼in) waste pipe

Basin details

Pedestal basins

Run pipework up to and behind a pedestal. Fix the basin to the wall with screws. Some basins are attached to the pedestal with clips, or bonded with silicone sealant. Screw the pedestal to the floor.

Pressed-metal basin

When you fit taps to a pressed-metal basin, slip built-up 'top-hat' washers onto the tails to cover the shanks. The basin itself may be supplied with a rubber strip to seal the joint with the counter top.

Counter-top basin

A template is supplied for cutting the hole in the counter top to receive the basin. Run mastic around the edge to seal the basin, and clamp it with the fixings supplied.

Top-hat washer
These large washers cover the shank of a tap fitted to a pressed-metal basin.

SEE ALSO > Draining the system 8, Connecting pipes 21–2, 25, Fitting taps 31

Choosing a new bath

Whether your bathroom is cupboard-like or palatial, there's a huge range of baths available to suit every conceivable location – and pocket. Built-in or freestanding, materials include acrylic, glass-fibre, pressed steel, cast iron, or even stone, with a wide variety of shapes and sizes to make bath time a real experience.

Selecting a bath

Access to a bath
Allow a 1100 x 700mm (3ft 8in x 2ft 4in) space beside a bath so that it's possible to climb in and out safely, and for bathing younger members of the family.

● **Selecting taps for a bath**
In design and style, bath taps are identical to basin taps; but they are proportionally larger, with 22mm (³⁄₄ in) tails. Some bath mixers are designed to supply water to a sprayhead, either mounted telephone-style on the mixer itself or hung from a bracket mounted on a wall above the bath.

Nowadays, the majority of baths are made from acrylic or glass-reinforced plastic – light and relatively cheap materials offering a wide variety of shapes. The surfaces are warm to the touch, and some freestanding models are 'double glazed' with an insulated twin-skin construction to help retain heat. The surfaces are vulnerable to abrasive cleaners and bleach, however, and care needs to be taken when using a blowtorch nearby for plumbing.

With their enamelled surfaces, pressed-steel and cast-iron baths are much tougher. They are heavy and it will take two people to carry a steel bath, and perhaps more for a cast-iron bath. Resin/stone composite baths are becoming more popular and are warm and practical.

Baths can be built in, panelled, sunken or freestanding. Corner baths can allow a bigger bath in a smaller space, while for very small bathrooms there are compact units just 1200mm (3ft 11in) long. There are also side-opening baths for those who have mobility problems.

Many baths can now be supplied fitted with a spa facility, where a powerful pump circulates water through a series of nozzles fitted in the sides of the bath – DIY kits are available to convert existing baths.

Rectangular bath
A standard rectangular bath is still the most popular and economical design. Baths vary in size from 1.5 to 1.8m (5 to 6ft) in length, with a choice of widths from 700 to 800mm (2ft 4in to 2ft 8in).

Corner bath
A corner bath occupies more floor area than a rectangular bath of the same capacity, but because it is fitted across the corner it may take up less wall space. A corner bath usually provides some shelf space for essential toiletries.

Round bath
A round bath is likely to be impractical in

most bathrooms – but if you are converting a spare bedroom, you may decide to make the bath a feature of the interior design as well as a practical appliance.

Freestanding bath
The Victorian roll-top bath is the best example of this style, though there are many modern variations on the theme.

An acrylic bath

Plastic can be moulded into attractive shapes

Corner baths make good use of space

Roll top baths are still popular

Supporting a frame-mounted plastic bath

A frame with adjustable feet is supplied to cradle a flexible plastic bath. The parts need to be assembled before the bath is fitted into place. To avoid any movement, it's important to ensure the feet are level and resting on a firm surface.

Assembling the cradle
Turn a bath onto its rim to fit the cradle.

If space is tight you can fit a compact bath

Illuminated bath panels give a modernist look

SEE ALSO > Selecting taps 30

Plumbing a bath

Once a bath is fitted close to the wall, it can be difficult to make the joints and connections – so fit the taps, overflow and trap before you push the new bath into position. Set the adjustable feet to raise the rim of the bath to the required height, and check it for level along its length and width. If the bath has small feet, cut two boards to go under them to spread the point load over a wider area.

Removing and installing

Turn off the water supply before you drain the system.

Removing an old bath
Have a shallow bowl ready to catch any trapped water, then use a hacksaw to cut through the old pipes. The overflow pipe from an old bath will almost certainly exit through the wall, so saw through the overflow at the same time.

If the bath has adjustable feet, lower them and then push down on the bath to break the mastic seal between the bathroom walls and the rim. Pull the bath away from the walls.

If a cast-iron bath is beyond restoration, it is easier to break it up *in situ* and carry it out in pieces. Drape a dust sheet over the bath; then, wearing gloves, goggles and ear protectors, smash it with a heavy club hammer.

Installing a new bath
Either run new 22mm (¾in) supply pipes or attach spurs to the existing ones, ready for connection to the flexible pipes already fitted on the bath taps.

Slide your new bath into position and adjust the height of the feet with a spanner. Use a spirit level to check that the rim is horizontal. Adjust the flexible tap pipes and join them to the supply pipes. Connect a 40mm (1½in) waste pipe to the trap and run it to the external hopper or soil stack, as for a washbasin.

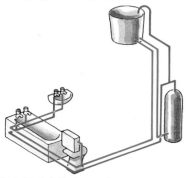
Typical tank-fed bathroom pipe runs
Red: Hot water. **Blue:** Cold water.

Fitting the taps

Fit individual hot and cold taps as for a washbasin. Fitting a mixer tap is a similar procedure, but some mixers are supplied with a long sealing gasket that slips over both tails. Lower the tails through the holes in the rim, then slip top-hat washers onto them and tighten both back-nuts to clamp the mixer securely to the bath.

Fit a flexible 22mm (¾in) copper pipe onto each tail. These flexible pipes allow for the easy adjustment that will be necessary if the joints are slightly misaligned. Alternatively, attach short lengths of standard 22mm (¾in) copper or plastic pipe with tap connectors, in preparation for jointing to the pipe run.

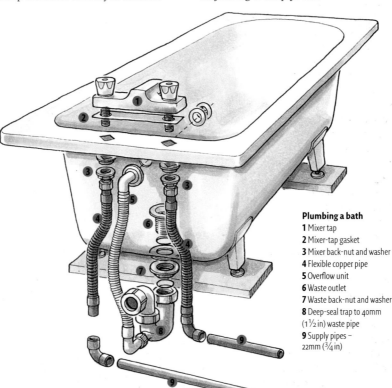

Plumbing a bath
1 Mixer tap
2 Mixer-tap gasket
3 Mixer back-nut and washer
4 Flexible copper pipe
5 Overflow unit
6 Waste outlet
7 Waste back-nut and washer
8 Deep-seal trap to 40mm (1½in) waste pipe
9 Supply pipes – 22mm (¾in)

Waste/overflow units
A flexible tube takes any overflow water to the trap.

Compression unit
Runs to the cleaning eye on the trap.

Banjo unit
Slips over the tail of the waste outlet.

WC and bath overflow
Overflow from a WC joins the bath unit.

Shallow-seal trap
Use this type of trap when space is limited. It must discharge to a yard gully or hopper, and not to a soil stack.

Fitting waste and overflow

With a combined waste and overflow, a flexible plastic hose takes water from the overflow outlet at the foot of the bath to the waste outlet or trap. If you use a 'banjo' unit, you must fit the overflow before the trap; but the flexible pipe of a compression-fitting unit connects to the trap itself (see above).

Put silicone sealant under the rim of the waste outlet, or fit a circular rubber seal. Before inserting its tail into the hole in the bottom of the bath, seal the thread with

PTFE tape. On the underside, add a plastic washer; then tighten the large back-nut, bedding the outlet down onto the sealant or the rubber seal. Wipe off excess sealant.

Connect the bath trap (see above) to the tail of the waste outlet and fit the banjo overflow unit at the same time.

Pass the threaded boss of the overflow hose through the hole at the foot of the bath. Slip a washer seal over the boss, then use a pair of pliers to screw the overflow outlet grille on.

SEE ALSO > Draining the system 8, Connecting pipes 21–2, 365, Fitting taps 31

Replacing a WC suite

Replacing an old WC with a modern suite is a relatively straightforward procedure, provided you can connect it to the existing branch of the soil pipe. However, if you are going to move a WC, or perhaps install a second one in another part of your home, you will have to connect to the main soil pipe itself or run the waste into the underground drainage system. In either case, hire a professional plumber to make these connections.

Cisterns

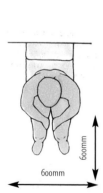

From antique-style high-level cisterns to discreet close-coupled or concealed models, the choice is so wide that you're bound to find one to suit your requirements. Before buying, make sure the equipment carries the British Standard 'Kite mark' or complies with equivalent EC standards.

High-level cistern

If you simply want to replace an old-fashioned high-level cistern without having to modify the pipework, comparable cisterns are still available from plumbers' merchants.

Standard low-level cistern

Many people prefer a cistern mounted on the wall just above the WC pan. A short flush pipe from the base of the cistern connects to the flushing horn on the rear of the pan, while inlet and overflow pipes can be fitted to either side of the cistern. Most low-level cisterns are manufactured from the same vitreous china as the WC pan.

Compact low-level cistern

Where space is limited, use a plastic cistern, which is only 114mm (4½in) from front to back.

Concealed cistern

A low-level cistern can be completely concealed behind panelling. The supply and overflow connections are identical to those of other types of cistern, but the flushing lever is mounted on the face of the panel. These plastic cisterns are utilitarian in character, with no concession to fashion or style, and are therefore relatively inexpensive. Don't forget that you will need to provide access for servicing.

Close-coupled cisterns

A close-coupled cistern is bolted directly to the pan, forming an integral unit. Both the inlet and overflow connections are made at the base of the cistern. An internal standpipe rises vertically from the overflow connection with the pan to protrude above the level of the water.

Choosing a WC cistern

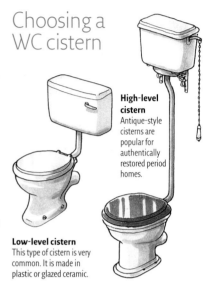

High-level cistern
Antique-style cisterns are popular for authentically restored period homes.

Low-level cistern
This type of cistern is very common. It is made in plastic or glazed ceramic.

Compact cistern
Very slim plastic cistern, for use where space is limited.

Concealed cistern
Plastic cistern for hiding behind panelling.

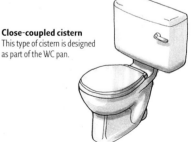

Close-coupled cistern
This type of cistern is designed as part of the WC pan.

WC pans

Space for a WC
You will need to allow a space at least 600mm (2ft) square in front of the pan.

600mm
600mm

When visiting a showroom, you are confronted with many apparently different WC pans to choose from, but in fact there are usually only two variations on the washdown pan – those where the cistern is connected by a pipe, and close-coupled units where the cistern bolts on directly.

Your bathroom may have an older, siphonic type of WC, but these are now obsolete due to their higher water usage.

WC pans are now made to reduce the amount of water needed to flush them. This is a legal requirement for new installations, but it is still possible and legal to buy an older-style pan with a higher water content.

New pans use the same trap seal depth and the waste pipe bore; they reduce the water content by having a shallow section rather like the shallow end of a swimming pool.

Washdown pans

Washdown pans work by displacement of waste by fresh water falling from the cistern. They are inherently more reliable than siphonic pans, but make considerably more noise when flushed.

Floor or wall exit?

When replacing a WC pan, check to see whether the new one needs to have a floor-exit or wall-exit trap.

Floor-exit trap
S-traps are connected to a soil pipe that is then passed through the floor.

Wall-exit trap
The outlet from a P-trap connects to a soil-pipe branch located behind the pan.

SEE ALSO > Cisterns 24–5

Removing an old WC

Cut off the water supply, then flush the cistern to empty it. If you are merely renewing a cistern, you will have to disconnect the supply and overflow pipes with a wrench and loosen the large nut connecting the flush pipe to the base of the cistern. These connections are often corroded and painted – so if you intend to replace the entire suite it is easier to hacksaw through the pipes close to the connections.

Choosing a WC pan

Washdown pan
The most common WC pan, with a simple trap filled with water.

Close-coupled pan
This washdown pan is neater than a pan with a separate cistern. The cistern bolts onto the back of the pan. P- or S-trap versions are available, though many are only made with a horizontal P-trap that can be connected to the soil pipe in the floor with a right-angled pan connector.

Wall-hung pan
A wall-mounted pan, connected to a concealed cistern, leaves the floor clear for cleaning. Unless it is built into the masonry, the pan is supported by a metal bracket/stand.

Removing the old pan and cistern

Remove the fixing screws through the back of the cistern, or lift it off its support brackets and remove them. Lever the brackets off the wall with a crowbar if necessary.

Cut the overflow pipe from the wall with a cold chisel. Repair the plaster when you decorate the bathroom.

If the pan is screwed to a wooden floor, it will probably have a P-trap connected to a nearly horizontal branch soil pipe. Remove the pan's floor-fixing screws and scrape out the old putty around the pipe joint. Attempt to free the pan by pulling it towards you while rocking it slightly from side to side.

If the joint is fixed firmly, smash the pan outlet just in front of the soil pipe with a club hammer (**1**). Protect your eyes with goggles. Stuff rags into the soil pipe to prevent debris falling into it, then chip out the remains of the pan outlet with a cold chisel (**2**). Work carefully, to preserve the soil pipe.

Smash an S-trap in the same way – and if the pan is cemented to a solid floor, drive a cold chisel under its base to break the seal. Chop out the broken fragments as before, and clean up the floor with a cold chisel.

If you crack the end of the soil pipe, cut it off square with a hired chain-link pipe cutter (**3**) or an angle grinder fitted with a masonry cutting disk. The cut-down pipe will still be able to accept a push-fit flexible connector, which should be lubricated before inserting into the pipe (**4**).

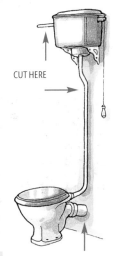

CUT HERE

CUT HERE

Removing an appliance
If fittings are corroded, remove the appliance by cutting through the flush pipe, overflow and pan outlet.

• **Lubricating connectors**
When installing plastic soil-pipe connectors, smear the surfaces lightly with a silicone lubricant.

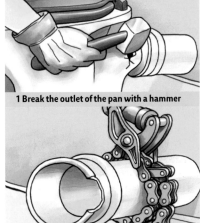

1 Break the outlet of the pan with a hammer

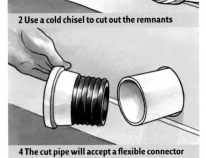

2 Use a cold chisel to cut out the remnants

3 Cut a damaged soil pipe with a hired cutter

4 The cut pipe will accept a flexible connector

Pan to soil-pipe connection

Before you install the new suite, choose a push-fit flexible connector to join the pan to the soil pipe. There are connectors to suit most situations, even when the two elements are slightly misaligned. You may need an angled connector to join a modern horizontal-outlet pan to an old P-trap branch pipe (see opposite).

When selecting a connector, make a note of the following dimensions: the external diameter of the pan outlet, the internal diameter of the soil pipe, and the distance between the outlet and the pipe when the pan is installed.

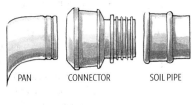

PAN   CONNECTOR   SOIL PIPE

OFF-SET   ANGLED   BENT

SEE ALSO > Turning off the water 8, Cisterns 24–5

Installing a new WC suite

Clean the floor and make good any damage before you begin to install a new WC suite. It's important to make sure that no debris enters the soil pipe while the WC is disconnected, so stuff it with an old cloth, or tape a carrier bag over it.

Fitting and plumbing the suite

Make up the new cistern with the siphon or valve and the inlet fill valve.

Not all cisterns come with instructions, but most have an illustration showing the position of the seals. Do not overtighten the nuts. Use enough pressure to press the rubber washers onto the ceramic, but not so much that the washers are squeezed out. If the washers get wet, it is best to dry them before tightening because the water lubricates them and helps them squeeze out from under the nut.

The pan is held onto the cistern with nuts and bolts. There will be a series of washers to seal the bolts and protect the ceramic surfaces. The golden rule is to avoid metal touching ceramic otherwise there's a risk of cracking.

The pan should be fixed to the floor with brass screws. Take care to align them correctly and avoid overtightening. The cistern should be fixed against the wall with screws that are cushioned with tap washers or a cistern fixing kit.

Alternatively, fix the cistern with a couple of dobs of silicone sealant – this reduces the possibility of the cistern cracking if there is any movement between the wall and floor.

On most cisterns the overflow now runs into the pan through an internal route. It is important to adjust the dump valve height so the cistern fills to the water line marked on the inside of the cistern.

In many cases this is set in the factory so you won't need to adjust it. If you are fitting a new dump valve, then you will often need to make an adjustment.

● **Fixing a new WC pan to the floor**
All manufacturers advise against the old-fashioned method of cementing a WC pan to a concrete floor. In fact, guarantees are usually invalidated if cement or a strong adhesive is used. If you can't screw the pan in place, just rely on the bed of silicone sealant to bond the pan to the floor.

Plumbing a WC
1 Flush push-button
2 Siphon
3 Cistern
4 Connecting plate
5 Foam washer
6 Flexible supply pipe
7 WC pan outlet
8 Flexible connector
9 Soil pipe

Typical pipe run
Red: Hot water
Blue: Cold Water

Small-bore waste systems

The siting of a WC is normally limited by the need to use a conventional 100mm (4in) soil pipe and to provide sufficient fall to discharge the waste into the soil stack. By using an electrically driven pump and shredder unit, you can discharge WC waste through a 22mm (¾in) pipe up to 50m (55yd) away from the stack. The shredder will even pump vertically, to a maximum height of about 4m (12ft).

You can run the small-bore pipework through the narrow space between a floor and ceiling. Consequently, a WC can be installed as part of an en-suite bathroom, in a basement, even under the stairs, so long as the space is adequately ventilated.

The unit is designed to accept any conventional P-trap WC pan. It is activated by flushing the cistern, and switches off about 18 seconds later. It must be wired to a fused connection unit – via a suitable flex outlet if it is installed in a bathroom.

The waste pipe can be connected to the soil stack using a 32mm (1¼in) pipe boss, provided the manufacturer supplies a 22 to 32mm (¾ to 1¼in) adaptor. A WC waste pipe must be connected to the soil stack at least 200mm (8in) above or below any other waste connections.

It's wise to check that these systems are approved by your local water supplier.

Small-bore waste system for a WC
The shredding unit fits neatly behind a P-trap WC pan. When fitted in a bathroom, the unit must be wired to a flex outlet. Otherwise, it can be connected to a fused connection unit.

SEE ALSO > Building Regulations on electrical wiring 70, Fused connection units 78, Flex outlet 79

Choosing a shower

All showers, except for the most powerful, use less water than required for filling a bath. And because showering is generally quicker than taking a bath, it helps to alleviate the morning queue for the bathroom. For even greater convenience, install a second shower somewhere else in the house – this is one of those improvements that really does add value to your home. Improvements in technology have made available a variety of powerful, controllable showers. However, many appliances are superficially similar in appearance, so it's important to read the manufacturers' literature carefully before you opt for a particular model.

Pressure and flow

When choosing a shower, bear in mind that pressure and flow are not the same thing. For example, an instantaneous electric shower delivers water at high mains pressure, but a relatively low flow rate is necessary to allow the water to heat up as it passes through the shower unit.

A conventional gravity-fed system delivers hot water from a storage cylinder under low pressure, but often has a high flow rate when measured in litres per minute. Adding a pump to this type of system can increase both pressure and flow rate. Alter the flow and pressure ratio by fitting an adjustable showerhead with a choice of spray patterns, from needle jets to a gentle 'champagne' cascade .

This showerhead provides a choice of spray patterns

Gravity-fed showers

In most homes cold water is stored in a tank in the loft and fed to a hot-water cylinder at a lower level. Both the hot-water and cold-water pressures are determined by the height (or 'head') of the cold-water storage tank above the shower. A minimum of at least 1 metre (3ft) head should give reasonable flow rate and pressure. If flow and pressure are insufficient for a good shower, you could improve the situation by raising the tank or fitting a pump in the system.

Mains-pressure showers

Some types of shower are fed directly from the mains: one of the simplest to install from a plumbing point of view is an instantaneous electric shower, which needs a dedicated electricity supply.

Another alternative is to install a thermal-store cylinder. Mains-pressure water passes through a rapid heat exchanger inside the cylinder (see right). Yet another option is to store hot water in

an unvented cylinder – which will supply high-pressure water to a shower without the need for a booster pump.

Nowadays showers are often supplied from combination boilers, though these often need to run at full flow to keep the boiler firing properly. Before buying a shower, check with the manufacturer of your boiler to ascertain whether there's likely to be a problem.

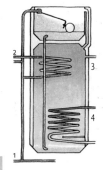

Thermal-store cylinder
Mains-fed water passes through a rapid heat exchanger on its way to the shower.
1 Mains feed
2 To shower
3 Other outlets
4 Boiler connections

Shower enclosures
If space permits, choose an enclosed shower cubicle (centre left). However, there are a number of screens and plumbing options, which make an over-the-bath shower almost as efficient.

Drainage

Draining the used water away from a shower can be more of a problem than running the supply.

If it is not possible to run the waste pipe between the floor joists or along a wall, then you may have to consider relocating the shower. In some situations it may be necessary to raise the shower tray on a plinth in order to gain enough height for the waste pipe to fall (slope) towards the drain. Another way to overcome the problem is to install a special pump to take the waste water away from the shower.

Shower traps

When running the waste pipe to an outside hopper, you can fit a conventional

trap – but these are relatively large, creating problems when installing the shower tray.

You could cut a hole in the floor, or substitute either a smaller, shallow-seal or compact trap that includes a removable grid and dip tube for easy cleaning. Another possibility is to fit a running trap in the waste pipe at a convenient location, or install a self-sealing valve in the pipe.

A shower trap that is connected to a soil stack must have a water seal not less than 50mm (2in) deep. The best solution is to fit a compact trap, which is shallow enough to fit under most modern shower trays, but is designed to provide the necessary water seal. Alternatively, fit either a running trap or a self-sealing valve.

Wet rooms
Wet rooms do away with shower trays and have drainage through the floor. They create a clean, uncluttered look, but need careful waterproofing to avoid problems with leaks

SEE ALSO > Shower mixers and sprayheads 40, Booster pumps 44, Thermal-store cylinders 53, Unvented cylinders 53

Shower mixers and sprayheads

Installing an independent shower cubicle with its own supply and waste systems requires some prior experience of plumbing – but if you use an existing bath as a shower tray, then fitting a shower unit can involve little more than replacing the taps.

Bath/shower mixers

This type of shower is the simplest to install. It is connected to the existing 22mm (¾in) hot and cold pipes in the same way as a standard bath mixer, and the bath's waste system takes care of the drainage. Once you have obtained the right temperature at the spout by adjusting the hot and cold valves, you lift a button on the mixer to divert the water to the sprayhead via a flexible hose. The sprayhead can be hung from a wall-mounted bracket to provide a conventional shower, or hand-held for washing hair.

Because the supply pipes for this type of shower are part of the overall house system, it's impossible to guard against fluctuating pressure – and potentially scalding temperatures – unless the mixer is fitted with a thermostatic valve. An unregulated shower could be a real hazard if there are very young or elderly residents. Installing a pressure-equalizing valve in the pipework will add convenience and safety – if the pressure is insufficient, fit a booster pump.

Don't fit a bath/shower mixer unless both the hot and cold water is under the same pressure, either high or low. This type of shower has controls that are uncomfortably low to reach.

Manual shower mixers
A manual shower mixer can be located over a bath or in a separate shower cubicle. Manual mixers require their own independent hot and cold supply.

Simple versions are available with individual hot and cold valves, but most manual shower mixers have a single control that regulates the flow and temperature of the water. Single-lever ceramic-disc mixers operate exceptionally smoothly and, having few moving parts, are less prone to hard-water scaling.

You can choose a surface-mounted unit, or a nearly flush mixer with the pipework, connections and shower mechanism all concealed in the wall.

Thermostatic mixers

A thermostatic shower mixer is similar in design to a manual mixer, but it has an extra control incorporated to preset the water temperature. If the flow rate drops on either the hot or cold supply, a thermostatic valve rapidly compensates by reducing the flow on the other side. This is primarily a safety measure, to prevent the shower user being scalded should someone run a cold tap elsewhere in the house. A thermostatic shower should be supplied by means of branch pipes from the bathroom plumbing – but try to join them as near as possible to the cold tank and hot cylinder. The mixer can't raise the pressure of the supply, so you still need a booster pump if the pressure is low.

Thermostatic mixer mechanisms are usually based on wax-filled cartridges or bimetallic strips. Brand-new thermostatic valves respond extremely quickly to changes of temperature, but the rate slows down as scale gradually builds up inside the mixer. Even when new, reaction time will be slower if the mixer is expected to cope with exceptionally hot water (above 65°C/149°F). At such high temperatures the hot-water ports are almost fully closed and the cold ones almost wide open, so there is little margin for more adjustment.

The majority of thermostatic mixers can be used with the existing gravity-fed hot and cold supply, but it may be necessary to fit a booster pump. Check the manufacturer's literature – some showers don't perform well at low pressures.

Sprayheads

There's a wide range of sprayheads available and many are adjustable to offer a variety of spray patterns. If you're thinking of upgrading an existing shower by installing an electric pump, it's worth finding out whether you can also replace the sprayhead.

In addition to the standard shower spray, a simple adjustment is all that is needed to produce an invigorating, pulsing jet to wake you up in the morning or a soft bubbly stream that is ideal for small children. Some sprayheads can also be adjusted to deliver a very light spray while you soap yourself or apply shampoo.

Rubber nozzles
Rub the the rubber nozzles on a modern sprayhead occasionally with your hand to help to shift lime scale accumulation.

Cleaning a sprayhead
Gradually, accumulation of lime scale blocks the holes in the sprayhead, and eventually this affects the performance of your shower. The harder the water, the more often it will need cleaning.

Remove the entire sprayhead from its hose or unscrew the perforated plate from the showerhead. Leave the sprayhead or plate to soak in a proprietary descalant until the scale has dissolved, then rinse thoroughly under running cold water.

Before you reattach the sprayhead or plate, turn on the shower to flush any loose scale deposits from the pipework.

Dual-lever mixer
With this type of mixer, one lever controls flow and the other temperature.

Bath/shower mixer
Fit this type of shower unit like an ordinary bath mixer.

Thermostatic mixer
This unit prevents excessive fluctuations in water temperature.

SEE ALSO > Choosing a shower 39, Fitting a bath mixer 35

Instantaneous showers

An instantaneous electric shower is designed specifically for connection to the mains water supply, using a single 15mm (½in) branch pipe from the rising main. A non-return valve must be fitted close to the unit.

Incoming water is heated within the unit, so there is no separate hot-water supply to balance. The shower is thermostatically controlled to prevent fluctuations in pressure affecting the water temperature and will switch off completely if there is a serious pressure failure. Some units have a shut-down facility: when you switch off, the water continues to flow for a little while to flush any hot water out of the pipework. This ensures that someone stepping into the cubicle immediately after another user isn't subjected to an unexpectedly hot start to their shower.

The electrical circuit

An instantaneous shower requires a separate circuit from the consumer unit. A ceiling-mounted double-pole switch is connected to the circuit to turn the appliance on and off.

Surface-mounted or concealed

With most instantaneous showers, all plumbing and electrical connections are contained in a single mixer cabinet that is mounted in the shower cubicle or over the bath. However, you can buy showers with a slim, flush-fitting control panel that is connected via a low voltage cable to a power pack installed out of sight – for example, under the bath behind a screw-fixed panel.

Fit a stopcock or miniature isolating valve in the supply pipe to allow the shower to be serviced.

Pump-assisted showers

The pump-assisted 'power' shower is perhaps most people's concept of the ideal shower. The pump delivers water at a constant pressure and flow rate, eliminating the need for the minimum pressure normally required for a gravity-fed shower.

Most power showers need a head of about 75 to 225mm (3 to 9in) to activate the pump when the mixer control is turned on. A pump can be used to boost the pressure and flow rate of stored hot and cold water, but not mains-fed water.

Ideally, the cold supply should be taken directly from the storage tank – not from branch pipes that feed other taps and appliances. The hot-water supply can be connected to the cylinder by means of a Surrey or Essex flange, which helps to eliminate the tendency for the pump to suck in air from the vent pipe.

If the water is heated by an electric immersion heater, make sure the cylinder is fed by a dedicated cold feed and that the cold-feed gate valve is fully open. This is to prevent the top of the cylinder running dry and perhaps burning out the heater. If the cylinder is heated from a boiler, make sure the water temperature is controlled by a thermostat. If the water is too hot, the shower could splutter.

Many power showers resemble instantaneous units, with the mixer controls and pump enclosed in a waterproof casing mounted on the shower cubicle wall.

Other pumps are designed for remote installation in the pipes feeding the shower mixer. These freestanding pumps can also be used to improve the performance of an existing installation. They're usually located next to the hot-water cylinder in an airing cupboard – as low as possible, so that the pump remains full of water. There are also pumps that are designed to perform satisfactorily when mounted at a high level – even in a loft. In such situations, a single-impeller pump is best.

Water Regulations

If the shower is mounted in such a way that the sprayhead could dangle below the rim of the bath or shower tray, you have to fit double-seal non-return valves in the supply pipes to prevent dirty water being siphoned back into the system. Most shower sets come with a bracket to prevent the hose reaching that far.

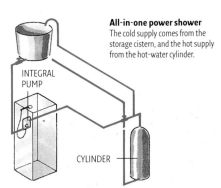

All-in-one power shower
The cold supply comes from the storage cistern, and the hot supply from the hot-water cylinder.

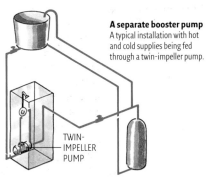

A separate booster pump
A typical installation with hot and cold supplies being fed through a twin-impeller pump.

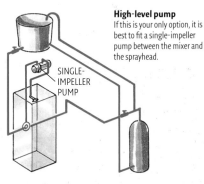

High-level pump
If this is your only option, it is best to fit a single-impeller pump between the mixer and the sprayhead.

Computer-controlled showers

Computerized showers allow for the precise selection of temperature and flow rates, using a touch-sensitive control panel. Most panels also include a memory program, so that each member of a family can select their own preprogrammed ideal shower.

These showers have real advantages for the disabled and for elderly people since they are exceptionally easy to operate – and the control panel can even be mounted outside the cubicle, so that it's possible to warm up the shower before you get in, or operate it for someone else.

Touch-sensitive computerized panel

SEE ALSO > Building Regulations on electrical wiring 70, Wiring electric showers 81

Building a shower enclosure

The simplest way to acquire a shower cubicle is to install a factory-assembled cabinet, complete with tray and mixer, together with waterproof doors or a curtain to contain the spray from the sprayhead. Once you have run supply pipes and drainage, the installation is complete. However, factory-built cabinets are expensive and there is an alternative – to construct a purpose-made shower cubicle to fit the allocated space.

Choosing the site

When deciding upon the location of your shower, consider whether you can use the existing walls – or do you need new partitions to enclose the cubicle?

Freestanding

You can place the shower tray against a flat wall and either construct a stud partition on each side or surround the tray with a proprietary enclosure.

Corner site

If you position the tray in a corner of a room, then two sides of the cubicle are ready-made. Run a curtain around the tray or install a corner-entry enclosure with sliding doors. Alternatively, build a fixed side wall yourself and put either a door or a curtain across the entrance.

Built-in cupboards

To incorporate a shower cubicle unobtrusively in a bedroom, place it in a corner, as described above, then construct a built-in wardrobe between the shower and the opposite wall.

Proprietary shower enclosure

Proprietary unit
A typical kit includes a plastic corner pillar that conceals the plumbing. The kit comes complete with shower set, tray and enclosure.

Hiding the plumbing

One solution for concealing the pipes is to install a proprietary shower cubicle with a plastic pillar in the corner that is designed to hide the plumbing and house the mixer and adjustable sprayhead (see left).

If you erect a stud partition, then you can run the pipework between the studs. Screw exterior-grade plywood or cement-based wallboard on the inside of the frame for a tiled finish, or use prefinished uPVC bathroom wall panelling.

Mount the shower mixer and sprayhead. Finish the inside with ceramic tiles, then seal the shower tray joints with silicone sealant. It's easier to connect the plumbing to the shower mixer before you enclose the outside of the partition.

Cut decorative wall panelling to size, then fix using screws and the plastic corner profiles supplied. Finally, seal all joints with silicone sealant.

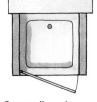
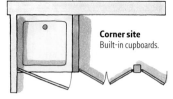
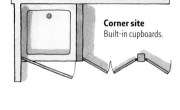

Freestanding unit
Two new partitions.

Freestanding unit
Proprietary enclosure.

Corner site
Enclosed by a curtain.

Corner site
Partition and curtain.

Corner site
Built-in cupboards.

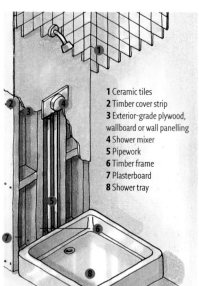

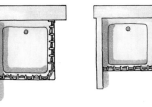

1 Ceramic tiles
2 Timber cover strip
3 Exterior-grade plywood, wallboard or wall panelling
4 Shower mixer
5 Pipework
6 Timber frame
7 Plasterboard
8 Shower tray

Running plumbing through a partition
Conceal pipework in a simple timber partition covered with ceramic tiles or panelling.

Shower trays

Shower trays are made from a variety of materials, but plastic trays are the most common. Lightweight ABS plastic trays tend to flex slightly in use, so it's particularly important to seal the edges carefully, using a good-quality silicone sealant (don't rely on grout). GRP trays are more substantial, while resin-bonded trays or ceramic trays are very solid.

The majority of shower trays are between 750 and 900mm (2ft 6in and 3ft) square. You can also buy trays that have a cut-off or rounded corner to save floor space. Larger rectangular trays provide more elbow room.

Most trays are designed to stand on a timber or masonry frame so that they lie about 150mm (6in) off the ground. Some have adjustable feet or a metal underframe helping to provide a fall for the waste pipe. Rigid types can be bedded in mortar. A plinth screwed across the front of the tray hides the underframe and plumbing, and provides access to the trap for servicing. Some shower trays are intended to be sunk, so that they are flush with the floor.

Enclosing a shower

A shower in a cubicle or over a bath needs to be provided with some means of preventing water spraying out onto the floor. Hanging a waterproof fabric curtain on curtain track or a tubular shower rail across the entrance is the simplest and cheapest method, but it is not really suitable for a power shower.

For a more satisfactory enclosure, use a metal-framed glass or plastic panelled unit. Hinged, sliding, or concertina doors operate within an adjustable frame fixed to the top edge of the tray and the side walls. Bed the lower track onto mastic to make a waterproof joint with the tray and, once you have completed the enclosure, run mastic between the framework and the surrounding walls.

SEE ALSO > Running and concealing pipework 28

Gravity-fed showers

Use the procedure below as a guide to the stage-by-stage installation of a cubicle and conventional gravity-fed shower. Check the instructions for your shower unit for any special requirements.

Ideally, you should run an independent cold supply from the storage tank; and for the hot supply, take a branch pipe directly from the vent pipe above the hot-water cylinder. Fit isolating gate valves in both supplies. Use the methods described earlier in this chapter for fitting plastic or copper supply pipes and drainage, in conjunction with the manufacturer's recommendations.

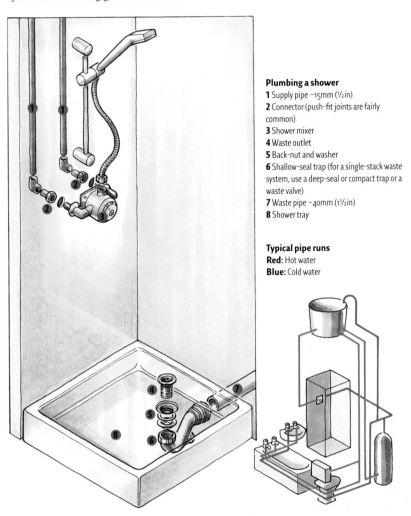

Plumbing a shower
1 Supply pipe –15mm (½in)
2 Connector (push-fit joints are fairly common)
3 Shower mixer
4 Waste outlet
5 Back-nut and washer
6 Shallow-seal trap (for a single-stack waste system, use a deep-seal or compact trap or a waste valve)
7 Waste pipe –40mm (1½in)
8 Shower tray

Typical pipe runs
Red: Hot water
Blue: Cold water

Installing an electric shower

If you're installing an instantaneous shower in the cubicle, run both the 10mm² electrical supply cable and a single 15mm (½in) pipe from the rising main through the stud partition.

Fit a non-return valve and an isolating valve in the pipe. Drill two holes in the wall just behind the shower unit for the pipe and cable. Join a threaded or compression connector to the supply pipe, whichever is appropriate for the water inlet built into the shower unit.

Read the section in this book about wiring an instantaneous shower, especially the Building Regulations that relate to electrical wiring.

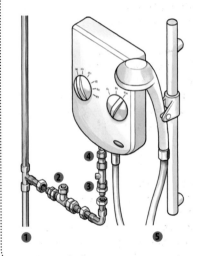

Plumbing an instantaneous shower
1 15mm (½in) pipe
2 Isolating valve
3 Non-return valve
4 Tap connector from rising main
5 Hose to sprayhead

Fit the waste outlet in the shower tray and connect a shallow-seal trap, as for a bath. Alternatively, fit a compact trap that has a removable grill for easy cleaning.

Install the tray and run a 40mm (1½in) waste pipe to a hopper or soil stack. In the latter case, you must use a deep-seal trap or a suitable compact trap. Alternatively, you can fit a running trap or a waste valve. Check with your Building Control Officer.

To enclose a shower situated in a corner (see opposite), construct a stud partition on one side and line the inner surface with plywood or wallboard.

Cut a hole in the board for a flush-fitting shower mixer; or drill holes for the supply pipes to a surface-mounted model. Tile the inside of the cubicle with ceramic tiles, using waterproof adhesive and grout.

Fit the shower mixer and sprayhead to the tiled surface. Connect the pipework and run it back to the point of connection with the water supplies. Fit an isolating valve to each of the supply pipes, then turn off the water and make the connections.

Fix the cover Attach the cover to protect the inner workings

Hose restraint A hose restraint stops the head from falling into the bath

Cleaning compact traps Compact shower traps have a dip tube for easy cleaning

SEE ALSO > Turning off the water 8, Connecting pipes 21–2, 25, Building Regulations on electrical wiring 70, Wiring electric showers 81

Installing power showers

If you're installing a brand-new power shower, it probably pays to opt for an all-in-one model with an integral pump. If you are merely unhappy with the performance of your existing shower, then it's much cheaper and more convenient to plumb in a separate pump.

Whichever system you choose, check that your cold-water storage tank is big enough – typically a minimum capacity of 115 litres (25 gallons). Some manufacturers also recommend a hot-water cylinder with a minimum 161 litres (35 gallons) capacity.

Both types of shower need an electrical supply to drive the pump. The pump is wired to a ring main by means of a switched fused connection unit installed outside the bathroom, or you can fit a ceiling-mounted double-pole switch inside the bathroom. The shower pump switches on automatically as soon as the shower valve is operated.

Fitting an all-in-one shower

To plumb a shower with an integral pump, you can run dedicated hot and cold supplies to the shower, as when fitting a gravity-fed shower. Alternatively, you can connect the hot-water supply directly to the cylinder by using a cylinder flange. An Essex flange is connected to the side of the cylinder (1); but to avoid cutting into the cylinder wall, fit a Surrey flange that screws into the vent-pipe connection on top of the cylinder (2). Fit gate valves in the hot and cold pipes, so you're able to isolate them for servicing.

All-in-one showers are prone to vibration: on a timber-frame wall this can create considerable noise. Isolate the unit by mounting it on rubber tap washers slid over the fixing screws.

All tiling and grouting needs to be completed before mounting the shower on the wall.

1 Essex flange　　**2 Surrey flange**

3 Unscrew vent-pipe　　**4 Attach hot supply**

Installing the shower

Drain the cold-water tank and drill a hole for a tank-connector fitting. Fit a gate valve close to the tank and run the pipe to the shower unit. Turn off the cold supply to the hot-water cylinder, and then open the hot taps in the bathroom to drain a small amount of water from the cylinder. Unscrew the vent-pipe connector (3) and catch any water left in the pipe with a towel.

Wrap PTFE tape around the threads of the Surrey flange, then screw it into the cylinder. Connect the original vent pipe to the top of the flange and run the hot supply for the shower from the side connection (4). Open the gate valves briefly to flush the pipes.

Run the electrical cable to the shower, ready for connection. (See BUILDING REGULATIONS ON ELECTRICAL WIRING.)

Mount the shower unit, using the screws provided and taking care not to bore into pipes or cable.

Connect the pipes to the unit and connect up the electrical cable to the terminal block inside the unit. Metal pipes must be bonded to earth.

Typical pipe runs
Red: Hot water
Blue: Cold water

FUSED CONNECTION UNIT

GATE VALVE
SHOWER UNIT

CYLINDER

Power shower with integral pump

Installing a booster pump

Fitting an electric pump can improve the performance of an existing shower. If you have access to the pipe running from the mixer to the sprayhead, you can install a single-impeller pump that boosts ready-mixed hot and cold water (1). If the pipework is embedded behind tiling, install a twin-impeller pump in the supply pipes before the mixer. The same twin-impeller pump can boost the supply to other outlets in the bathroom, too. (2).

Positioning the pump

Place the pump somewhere convenient for servicing, perhaps on the floor under the bath, behind a screw-fixed panel – but not where it will be splashed with water. Stand it on a foam mat or pads to reduce the noise from vibration, and don't screw it to the floor. Most pumps come with push-fit flexible connectors to join pipes to the pump and help to prevent vibration.

Connect up the pump to a switched fused connection unit (see top left). In use, the pump is activated automatically by flow switches.

The basic plumbing is identical to that described for installing an all-in-one shower. Flush the pipes before you switch on the pump.

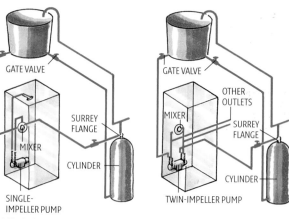

GATE VALVE

SURREY FLANGE

MIXER

CYLINDER

SINGLE-IMPELLER PUMP

1 Single-impeller pump
Boosts ready-mixed water.

GATE VALVE

OTHER OUTLETS

MIXER

SURREY FLANGE

CYLINDER

TWIN-IMPELLER PUMP

2 Twin-impeller pump
Can boost other outlets as well as a shower.

SEE ALSO > Turning off the water 8, Connecting pipes 21–2, 25, Storage tanks 51, Building Regulations on electrical wiring 70, Bathroom safety 72, Zones for bathrooms 73, Fused connection units 78

Plumbing a bidet

A bidet is primarily for washing the genitals and lower body, though it can double as a footbath. Because of the stringent requirements of the Water Regulations, installing some types can be an expensive and time-consuming procedure. However, simpler versions are plumbed in just like a washbasin.

Rim-supply bidet

This type delivers warm water to the basin via a hollow rim, making it warmer to sit on. A douche spray provides the water for washing from the bottom of the basin. A device in the centre of the mixer tap diverts water from the rim to the sprayhead.

Because the sprayhead is submerged when the basin is full, it's possible that dirty water could flow back into the supply pipes. To prevent any risk of contamination, the Water Regulations lay down strict rules about how the bidet is plumbed in. There are several methods allowed, but in general the appliance must have a dedicated cold water supply running directly from the storage tank, with no other connections to it. The hot-water supply must be independent and connected to the vent pipe above the cylinder. It's essential to check with your water supplier before installing a bidet.

Over-rim-supply bidet

This type of bidet is simply a low-level basin. It is fitted with individual hot and cold taps or a basin mixer, and has a built-in overflow running to the waste outlet in the basin. The disadvantage of an over-rim bidet is that the rim is cold to sit on.

Installing the bidet

When plumbing an over-rim-supply bidet, use exactly the same procedures, pipes and connectors described for plumbing a washbasin. Fit the taps, waste outlet and trap, then use a spirit level to position the bidet before fixing it to the floor with non-corrosive screws and rubber washers. Supply the hot and cold taps with branch pipes from the existing bathroom plumbing, and take the waste pipe to the hopper or stack.

When attaching the bidet set and trap to a rim-supply appliance, follow the manufacturer's instructions. Screw the bidet to the floor before running 15mm (½in) supply pipes and a 32mm (1¼in) waste according to the Water Regulations (see below). Connect the cold supply to the tank at the same level as the existing supply pipe.

Space for a bidet
When planning the position of a bidet, allow sufficient knee room on each side – about 700mm (2ft 4in) overall.

700mm

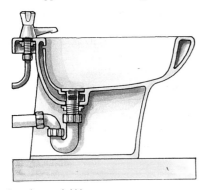

Over-rim-supply bidet
This type of bidet is simple to install. Follow the same procedure as for a washbasin.

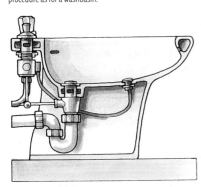

Rim-supply bidet
The installation of this type of bidet is complicated by the submerged douche spray. Independent plumbing is essential, as is a special mixer set to comply with the Water Regulations.

Plumbing an over-rim-supply bidet
1 Tap
2 Tap back-nut and washer
3 Tap connector
4 Supply pipe –15mm (½in)
5 Waste outlet
6 Waste back-nut and washer
7 Trap
8 Waste pipe –32mm (1¼in)

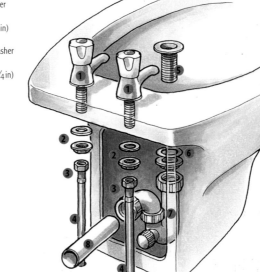

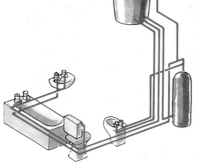

Over-rim-supply bidet
Typical pipe runs.
Red: Hot water
Blue: Cold water

SEE ALSO > Connecting pipes 21–2, 25, Washbasins 32–3, Taps 30–1, Tank supply 51

Kitchen sinks

If your ambition is to re-create a period-style kitchen, you may want a reproduction Butler or Belfast fire-clay sink with a separate teak or ceramic draining board. Alternatively, you could choose a stainless-steel sink top incorporating one or more bowls and drainer in a single pressing. You can match the colours in your kitchen with plastic, resin, ceramic, or one-piece moulded sink/worktop combinations in a wide variety of styles.

Choosing a kitchen sink

Choose the sink to make the best use of available space and to suit the style of your kitchen. If you don't have an automatic dishwasher, the kitchen sink must be large enough to cope with a considerable volume of washing-up (don't forget to allow for larger items, such as baking trays and oven racks). In addition, check that the bowl is deep enough to allow you to fill a bucket from the kitchen tap.

If space allows, select a unit with two bowls. If you plan to install a waste-disposal unit, one of the bowls will need to have a waste outlet of the appropriate size. Some sink units have a small bowl designed for waste disposal.

A double drainer is another useful feature; if there's no space, allow at least some room to the side of the bowl, to avoid mixing soiled and clean crockery.

One-piece sink tops are generally made to modular sizes to fit standard kitchen base units. However, many sinks are designed to be set into a continuous worktop – which offers greater flexibility in size, shape and, above all, positioning.

Sink units and taps

There's a wide range of kitchen sinks and taps available for the domestic market. Steel, enamel, resin, ceramic, double, single, plain, coloured – a huge choice confronts you when you are planning your kitchen. A cross section of popular sinks, accessories and taps is shown below to assist you in making your decision.

Double bowl with left-hand drainer

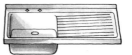

Single bowl with right-hand drainer

Inset double-bowl unit

Inset unit with waste-disposal bowl

Individual sink and drainer

Inset stainless-steel sink
This one-and-a-half bowl unit has provision for a waste-disposal unit and is designed to be inset into a worktop.

Belfast or Butler sink
A period style still popular. Made of enamelled fireclay, it is deep and practical. Ceramic drainers are available to match.

Kitchen taps

Except for being somewhat taller, kitchen taps are comparable in style to those used for washbasins. They also incorporate similar mechanisms and are fitted using the same methods.

A kitchen mixer, however, has an additional feature: drinking water is supplied to it from the rising main, whereas the hot water usually comes from the same storage cylinder that supplies all the other hot taps in the house. A sink mixer should have separate waterways to isolate the one supply from the other until

the water emerges from the spout; otherwise, you must have special check valves to prevent possible contamination of your drinking water.

If you are fitting a double-bowl sink, choose a mixer with a swivelling spout. Some sink mixers have a hot-rinse spray attachment for removing food scraps from crockery and saucepans.

Many mixer taps are supplied with small-bore copper or flexible tail pipes, which are joined to the supply pipes by a compression-joint or push-fit reducer.

Swivel mixer   **Pull-out spray**

Accessories for a kitchen sink

You can buy a variety of accessories to fit most kitchen sinks, including a hardwood or laminated-plastic chopping board that drops neatly into the rim of the bowl or drainer, and a selection of plastic-dipped wire baskets for rinsing vegetables or draining crockery.

Pump-action dispensers for soap and washing-up liquid rid the sink of plastic bottles and soap dishes.

Single-lever tap   **Pot filler**

SEE ALSO > Taps 30–31

Installing a sink

The trend for coloured sinks has given way to practical stainless steel. Another alternative is ceramic or Corian acrylic polymer, which incorporates the sink into a seamless worktop. This is favoured where hygiene is a critical issue since it does away with the crevices that can harbour bacteria.

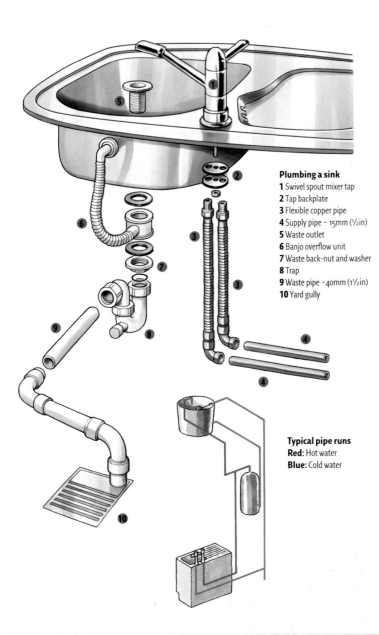

Plumbing a sink
1 Swivel spout mixer tap
2 Tap backplate
3 Flexible copper pipe
4 Supply pipe – 15mm (½in)
5 Waste outlet
6 Banjo overflow unit
7 Waste back-nut and washer
8 Trap
9 Waste pipe – 40mm (1½in)
10 Yard gully

Typical pipe runs
Red: Hot water
Blue: Cold water

Plumbing the sink

Fit the taps and the overflow/waste outlet to the new sink before you place the sink in position.

Turn off the water supply to the taps, then remove the old sink by dismantling the plumbing. Remove the old pipework unless you plan to adapt it.

Install the new sink on its base unit or worktop, using the fittings provided; then, if needed, seal the rim of the sink. Run a 15mm (½in) cold-water supply pipe from the rising main, and a branch pipe of the same size from the nearest hot-water pipe. Fit miniature isolating valves in both of the supply pipes and connect them to the taps with flexible copper-tap connectors.

Fit the trap and run a 40mm (1½in) waste pipe through the wall behind the base unit to the yard gully. According to current Water Regulations, the pipe has to pass through the grid covering the gully (see left) but must stop short of the water in the gully trap. You can adapt an existing grid quite easily by cutting out one corner with a sharp hacksaw.

Anti-siphon trap
If your trap gurgles as the sink empties, you could replace it with an anti-siphon trap. This draws in air to break the vacuum in the waste pipe. You can also get T-fittings that do the same job – these are placed just downstream of the existing trap.

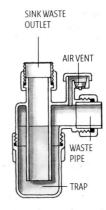

SINK WASTE OUTLET
AIR VENT
WASTE PIPE
TRAP

Kitchen sink materials

Plastic sinks are available in a wide range of styles and colours

Stainless steel is the most popular material for kitchen sinks as it resists heat and damage, wears well, is relatively cheap and can be formed into a wide variety of shapes. Being thin, it can be noisy in use, so sound-deadening pads are often applied to the underside.

Tough, heat-resistant plastic (encompassing a range of moulded materials) sinks come in a range of colours. The thick material is quiet in use. Sinks made from composite materials – often a mix of stone and resins – are becoming more popular for their looks and resilience, while there is still a place for traditional ceramic.

SEE ALSO > Miniature isolating valve 8, Tap connectors 20, Connecting pipes 21–2, 25, Overflow/waste 35

Washing machines

Washing machines and dishwashers are commonplace household appliances these days and they are pretty much identical in the way that they are plumbed in. They are attached by flexible hoses to a dedicated waste pipe or to the waste from the kitchen sink.

Waste-disposal units

A waste-disposal unit provides a hygienic method of dealing with soft food scraps – reserving the kitchen wastebin for dry refuse and bones.

The unit houses an electric motor that drives steel cutters or discs which grind up the food scraps into a fine slurry to be washed into the yard gully or soil stack. A continuous-feed model is operated by a manual, commonly pneumatic, switch: scraps are then fed into it while the cold tap is running. To prevent the unit being switched on accidentally, a batch-feed model cannot be operated until a removable plug is inserted in the sink waste outlet.

Waste-disposal units are generally designed to fit an 89mm (3½in) outlet in the base of the sink bowl.

With a sink waste outlet and seal in position, clamp a retaining collar to the outlet from under the sink and bolt or clip the unit housing to the collar.

The waste outlet from the unit itself fits a standard sink trap (not a bottle trap) and waste pipe. If the waste pipe runs to a yard gully, make sure it passes through the covering grid.

Wire the unit to a switched fused connection unit mounted above the worktop, out of the reach of children.

Waste-disposal unit
Units differ in detail, but the illustration shows the components typically used to clamp a waste-disposal unit to a sink.
1 Sink waste outlet
2 Gasket
3 Back-up ring
4 Collar
5 Snap ring
6 Unit housing
7 Cutters
8 Waste outlet
9 Trap

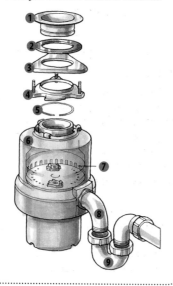

Many washing machines now have a cold fill only because lower wash temperatures mean that a hot fill is hardly ever required. Even if you have a hot and cold fill machine it can still be twinned into a single cold supply. This is often a more economical way of running the machine because you only heat the water required and do not have hot water standing in the supply pipes. In fact, a hot and cold fill machine might never get to use the hot water it draws because it fills from the water standing in the pipe, which is often cold. The water drawn from the heater or boiler is then simply sitting in the pipe going cold.

Water pressure

The machine's instructions should indicate what water pressure is required. If installed upstairs, make sure the drop from the storage tank to the machine is enough to provide the required pressure.

Downstairs there is rarely any problem with pressure if you can take the cold water from the mains supply. However, check with your water supplier if you want to connect more than one machine.

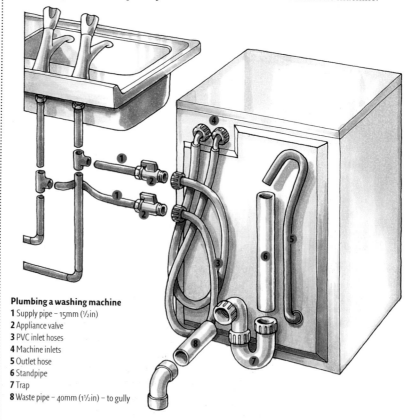

Plumbing a washing machine
1 Supply pipe – 15mm (½in)
2 Appliance valve
3 PVC inlet hoses
4 Machine inlets
5 Outlet hose
6 Standpipe
7 Trap
8 Waste pipe – 40mm (1½in) – to gully

Running the supply

Washing machines and dishwashers are supplied with PVC hoses to link the water inlets at the back of the appliance to service valves connected to the household plumbing. Using these valves, you can turn off the water when you need to service a machine, without having to disrupt the supply to the rest of the house. Select the type that provides the most practical method of connecting to the plumbing, depending on the location of the machine in relation to existing pipework.

SEE ALSO > Draining the system 8, Connecting pipes 21–2, 25, Storage tanks 51, Building Regulations on electrical wiring 70, Wiring kitchen appliances 79, Wiring a waste-disposal unit 79

Connecting to the supply

If you have to extend the plumbing to reach the machine, take branch pipes from the hot and cold pipes supplying the kitchen taps. Turn off the supply and drain down the pipe if possible, then cut through the existing pipes with a hacksaw (**1**) or pipe cutter, de-burr and clean the pipework, then insert a soldered or compression T-joint (**2**).

Add the branch pipes and terminate at a convenient position close to the machine. Then fit an appliance valve that has either a standard compression or push-fit joint for connecting to the pipework and a threaded outlet for the machine hose (**3**).

For branch pipes, use an in-line or elbow (right-angled) valve. A T-valve can be used where you have to break into an existing pipe running behind the machine. Self-cutting valves can also be used in this situation, but be aware that they may not provide enough flow for the appliance to function properly.

1 Cut into the supply pipes under the sink

2 Add a T-fitting for the branch pipe

3 Connect the hose to an appliance valve

Supplying drainage

The outlet hose from a dishwasher or washing machine must be connected to a waste system that will discharge the dirty water into either a yard gully or a single waste stack – not into a surface-water drain, where detergents could pollute rivers.

Standpipe and trap
The standard method, approved by all water suppliers, employs a vertical 40mm (1½in) plastic standpipe attached to a deep-seal trap (see opposite).

Most plumbing suppliers stock the standpipe, trap and wall fixings as a kit. The machine hose fits loosely into the open-ended pipe, so that dirty water won't be siphoned back into the machine. The machine manufacturer's instructions should tell you how to position the standpipe; in the absence of advice, ensure that the open end is at least 600mm (2ft) above the floor.

Cut a hole through the wall and run the waste pipe to a gully; or use a pipe boss to connect the waste to a drainage stack. Allow a minimum fall of 6mm (¼in) for every 300mm (1ft) of pipe run.

Draining to a sink trap
You can drain a washing machine to a sink

1 Sink trap with drainage spigot

2 In-line anti-siphon hose valve

trap that has a built-in spigot (**1**), but you should insert an in-line anti-siphon return valve in the machine's outlet hose. This is a small plastic device with a hose connector at each end (**2**). In order to drain a washing machine and dishwasher together, you will need a dual-spigot trap.

Appliance valves
Typical valves used to connect dishwashers and washing machines to the water supply.

Fitting an anti-siphon device

1 Clamp saddle over supply pipe

2 Screw the self-cutting valve in

3 Attach device and machine hose

Preventing a flood

Overflowing dishwashers and washing machines can cause a great deal of damage in just a few minutes – particularly if the appliance is plumbed into an upstairs flat and water is able to find its way through the building.

Air-inlet valves
Most overflows occur simply because the water backs up the waste pipe and spills out over the standpipe or sink.

A sealed waste system succeeds in overcoming this problem – since it does away with the air gap that allows the water to overflow. The anti-vacuum function is formed, instead, by a fitting that incorporates a small air-inlet valve, which stops the waste pipe siphoning the machine. The discharge hose from the machine is connected to the nozzle of the vent fitting, and a length of 40mm (1½in) waste pipe is inserted between the fitting and the washing machine trap under the sink.

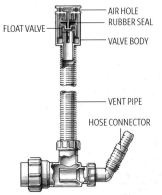

AIR HOLE
RUBBER SEAL
FLOAT VALVE
VALVE BODY
VENT PIPE
HOSE CONNECTOR

Anti-siphon devices
The standpipe-and-trap method of draining domestic appliances prevents back-siphonage by venting the pipe to the air, but there are other ways to deal with the problem. If an existing 32 or 40mm (1¼ or 1½in) waste pipe runs behind the machine, for example, you can attach a hose connector that incorporates a non-return valve to eliminate reverse flow. Connectors are available with short spigots (above), or can be attached to a standpipe.

SEE ALSO > Connecting pipes 21–2, 25, Cutting through a wall 29

Water softeners

Harmful impurities are removed from the water supply, but the amount of minerals absorbed from the ground determines whether water is hard or soft. Rocky terrain gives rise to surface-run water, which is naturally soft – but where water runs through the ground, rather than over it, the higher mineral content produces hard water.

Installing a water softener

Water softener
A domestic unit, which fits neatly beneath the worktop, requires topping up with salt.

Installing a water softener may appear to be fairly complicated since it involves a great deal of joint making – both to fit the valves and branch pipes that supply and bypass the softener and to include the fittings that are necessary to comply with the Water Regulations.

The bypass assembly allows for the unit to be isolated for servicing while maintaining the supply of water to the rest of the house. In addition, you must install a branch pipe before the assembly, in order to supply unsoftened drinking water to the kitchen sink. Supply your garden tap from the same pipe – there's no need to waste softened water on the garden.

Install a non-return valve in the system, to prevent the reverse flow of salty water. A pressure-reducing valve may also be required (check with your water supplier). You will need a draincock, in order to empty the rising main. Some manufacturers supply an installation kit that includes all the necessary equipment. You will have to provide drainage in the form of a standpipe and trap, as for a washing machine.

Wire the water softener to a switched 3amp fused connection unit.

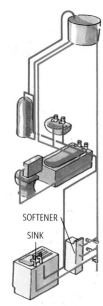

Typical pipe runs
A domestic system incorporating a softener.
Red: Hot water
Blue: Cold water

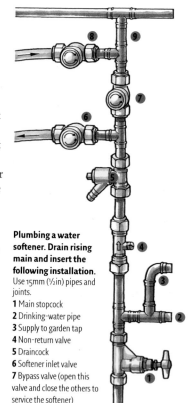

Plumbing a water softener. Drain rising main and insert the following installation.
Use 15mm (½in) pipes and joints.
1 Main stopcock
2 Drinking-water pipe
3 Supply to garden tap
4 Non-return valve
5 Draincock
6 Softener inlet valve
7 Bypass valve (open this valve and close the others to service the softener)
8 Softener return valve
9 Rising main

Domestic softeners

Water softeners work on the principle of ion exchange. The incoming water flows through a compartment containing a synthetic resin that absorbs scale-forming calcium and magnesium ions and releases sodium ions in their place.

After a period of about three or four days, the resin is unable to absorb any more mineral salts and the softener automatically flushes the compartment with a saline solution to regenerate the resin. Topping up with salt is required at intervals of perhaps two to three months. The softener is fitted with a timer so that you can programme regeneration when water consumption is at its lowest, usually

during the early hours of the morning.

The unit must be connected to the rising main at the point where the water supply enters the house. For this reason, domestic softeners are usually designed to fit under a kitchen worktop.

Lime scale effects

The more obvious consequences of hard water are the discoloration of baths and basins, blocked sprayheads, blemished stainless-steel surfaces and furred-up kettles. Most people resign themselves to living with these effects – but they can be reduced, or even eliminated altogether, by installing a water softener.

Outside taps

A bib tap situated on an outside wall is convenient for attaching a hose for a lawn sprinkler or for washing the car. To comply with the Water Regulations, a double-seal non-return (check) valve must be incorporated in the plumbing, to prevent contaminated water being drawn back into the system. Provide a means of shutting off the water and draining the pipework during winter, and keep the outside pipe run as short as possible. A self-cutting stoptap combining a non-return valve (below) simplifies the job.

A self-cutting stoptap incorporating a non-return valve (left) can be used to feed a garden tap (right)

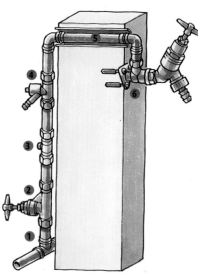

Pipes and fittings to supply a garden tap

Turn off and drain the mains supply. Fit a T-joint (1) to run the supply to the tap. Run a short length of pipe to a convenient position for another stopcock (2) or service valve, and for the non-return valve (3) if the tap doesn't include one, making sure that the arrows marked on both fittings point in the direction of flow. Fit a draincock (4) after this point. Run a pipe through the wall inside a length of plastic overflow (5), so that any leaks will be detected quickly and will not soak the masonry. Wrap PTFE tape around the bib-tap thread, then screw it into a wall plate attached to the masonry outside (6).

SEE ALSO > Draining the system 8, Connecting pipes 21–2, 25, PTFE tape 22, Washing machines 48, Building Regulations on electrical wiring 70, Fused connection units 78, Wiring kitchen appliances 79

Storage tanks

The cold-water storage tank, normally situated in the roof space, supplies the hot-water cylinder and all the cold taps in the house, other than the one in the kitchen that is used for drinking water. An old house may still have a galvanized-steel tank, which will eventually corrode, and, although it's possible to patch it, it makes sense to replace it before a serious leak develops. A circular 227 litre (50 gallon) polythene tank is a popular replacement, because it can be folded to pass through a narrow hatch to the loft.

Removing an old tank

Switch off all water-heating appliances, then close the stopcock on the rising main. Drain the storage tank by opening the cold taps in the bathroom.

Bail out the remaining water in the bottom of the tank, then use a spanner to dismantle the fittings connecting the float valve, distribution pipes and overflow to the tank. Use penetrating oil if the fittings are stiff with corrosion. Don't worry about damaging the old float valve as it's probably not worth using again.

The tank was probably put in the loft house before the roof structure was completed, so it almost certainly won't pass through the hatch. It would be too much of a job to cut it up, so just pull it to one side and leave it in the loft.

Prepare a firm base for the new tank by nailing stout planks across the joists, or lay a platform made from plywood 18mm (¾in) thick.

Plumbing a new tank

Once the new tank is in place, you can set about connecting the numerous pipes and fittings that are required.

Fitting the float valve
A float valve shuts off the flow of water from the rising main when the tank is full. Cut a hole for the float valve 75mm (3in) below the top of the tank. Slip a plastic washer onto the tail of the valve and pass it through the hole. Slide the reinforcing plate onto the tail, followed by another washer and a fixing nut, then tighten the fitting with the aid of two spanners.

Screw a tap connector onto the float valve, ready for connecting to the 15mm (½in) rising main.

Connecting the distribution pipes
The 22mm (¾in) pipes running to the cylinder and cold taps are attached by means of tank connectors – threaded inlets with a compression fitting for the pipework. Drill a hole for each tank connector, about 50mm (2in) above the

bottom of the tank. Push the fittings through each hole, with one polythene washer on the inside. Wrap a couple of turns of PTFE tape around the threads, then fit the other washer. Screw the nut on, holding the tank connector to stop it turning. Don't overtighten the nut – or you will damage the washer, causing it to leak.

Take the opportunity to fit a gate valve to each distribution pipe, so you can cut off the supply of water without having to empty the tank.

Connecting the overflow
Drill a hole 25mm (1in) below the level of the float-valve inlet for the threaded connector of the overflow-pipe assembly. Pass the connector through the hole, fit a washer, and tighten its fixing nut on the inside of the tank. Fit the dip pipe and insect filter.

Attach a 22mm (¾in) plastic overflow pipe to the assembly. Run the pipe to the floor, then to the outside of the house, maintaining a continuous fall. The pipe must emerge in a conspicuous position, so that an overflow can be detected immediately. Clip the pipe to the roof timbers.

Modifying existing plumbing
Modify the rising main and distribution pipes to align with their fittings, then connect them with compression fittings. Don't use soldered joints near a plastic tank as there's a danger that heat could travel along the pipes and melt the tank. Clip all the pipework securely to the joists.

Open the main stopcock and check for leaks as the tank fills. Adjust the float arm to maintain a water level 25mm (1in) below the overflow outlet.

Adapt the vent pipe from the hot-water cylinder to pass through the hole in the lid. Finally, insulate the tank and pipework – but make sure there is no loft insulation under the cistern, as this will prevent warmth rising from below.

Tank cutters
Hire a tank cutter to bore holes in the tank for pipework. Some cutters are adjustable, so you can drill holes of different diameters. An alternative is to use a hole saw clamped to a drill bit.

Hole saw

Adjustable cutter

● **Bylaw 30 kits**
Make sure your new tank is supplied with a Bylaw 30 kit, to keep the water clean. This is a requirement of all water suppliers. The kit includes a close-fitting lid that excludes light and insects, and is fitted with a screened breather and a sleeved inlet for the vent pipe. In addition, there should be an overflow-pipe assembly that is screened to prevent insects crawling into the tank, a reinforcing plate to stiffen the cistern wall around the float valve, and an insulating jacket.

Plumbing a tank
1 Float valve
2 Reinforcing plate
3 Tap connector
4 Rising main
5 Tank connector
6 Gate valve
7 Distribution pipe – 22mm (¾in)
8 Pipe clip
9 Overflow-pipe assembly
10 Overflow pipe
11 Vent pipe

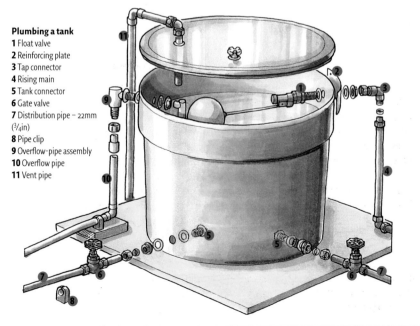

SEE ALSO > Insulation 22, Float valves 12–14, Adjusting a float arm 14, Gate valve 20, Tap connectors 20, Compression joints 22, Hot-water cylinders 52

Vented hot-water cylinders

In most houses, the hot water is heated and stored in a large copper cylinder situated in the airing cupboard. Cold water is fed to the base of the cylinder from the cold-water storage tank housed in the loft. As the water is heated, it rises to the top of the cylinder, where it is drawn off via a branch from the vent pipe to the hot taps. When the hot water is run off, it is replaced by cold water at the base of the cylinder, ready for heating.

Typical pipe runs
Red: Hot water
Blue: Cold water

Direct water heating by means of a boiler

Indirect water heating employs the central-heating boiler

The vent pipe itself runs back to the loft, where it passes through the lid of the cold-water storage tank, with its open end just above the level of the water. The vent pipe provides a safe escape route for air bubbles and steam, should the system overheat.

When water is heated, it expands. The vent pipe accommodates some of this, but much of the water is forced back up the cold-feed pipe into the cold storage tank.

Methods of heating water

There are two different methods of heating the water in a vented hot-water cylinder: either directly – usually by means of electric immersion heaters – or indirectly, by a heat exchanger connected to the central-heating system.

Direct heating
Water heating can be accomplished solely by means of electric immersion heaters – either a single-element or double-element heater is fitted in the top of the cylinder, or there may be two individual side-entry heaters.

An alternative is for the water to be heated in a boiler, the sole purpose of which is to provide hot water for the cylinder. A cold-water pipe runs from the base of the cylinder to the boiler, where the water is heated; and it then returns to the top half of the cylinder.

Both methods are known as direct systems. In practice, a boiler-heated cylinder is generally fitted with an immersion heater as well, so that hot water can be supplied independently during the summer, when using the boiler would make the room where it is situated uncomfortably warm.

Indirect heating
When a house is centrally heated with radiators fed by a boiler, the water in the cylinder is usually heated indirectly by a heat exchanger.

Hot water from the boiler passes through the exchanger (a coiled tube within the cylinder), where the heat is transmitted to the stored water. The heat exchanger is part of a completely self-contained system, which has its own feed-and-expansion tank (a small storage tank in the loft) to top up the system. An open-ended vent pipe terminates over the same small tank.

The whole system is known as the primary circuit, and the pipes running from and back to the boiler are known as the primary flow and return. An indirect system is often supplemented with an immersion heater, to provide hot water during the summer months.

Direct cylinder
1 Vent pipe
2 Hot-water branch pipe
3 Lower immersion heater (provides hot water using cheaper night-rate electricity)
4 Upper immersion heater (used for daytime top-up heating only)
5 Cold-feed pipe
6 Draincock

Indirect cylinder
1 Vent pipe
2 Back-up immersion heater
3 Flow from boiler
4 Heat exchanger
5 Return to boiler
6 Draincock
7 Cold feed from tank

Hot-water cylinders

The capacity of domestic cylinders normally ranges from about 114 litres (25 gallons) to 227 litres (50 gallons), although it is possible to obtain bigger cylinders. A cylinder with a capacity of between 182 and 227 litres (40 and 50 gallons) will store enough hot water to satisfy the needs of an average family for a whole day.

Some cylinders are made from thin, uninsulated copper and need to have a thick lagging jacket to reduce heat loss. However, for better performance use a Kite-marked factory-insulated cylinder that is precovered with a thick layer of foamed polyurethane.

Changing a cylinder

You may wish to replace an existing cylinder because it has sprung a leak, or because a larger one will allow you to take full advantage of cheap night-time electricity by storing more hot water. A simple replacement can sometimes be achieved without modifying the plumbing, but you'll have to adapt the pipework to fit a larger cylinder.

If you plan to install central heating at some point in the future, you can plumb in an indirect cylinder fitted with a double-element immersion heater and simply leave the heat-exchanging coil unconnected for the time being.

Place the new cylinder in position and check the existing pipework for alignment. Modify the pipes as need be, then make the connections, using PTFE tape to ensure that the threaded joints are watertight. Fit a draincock to the feed pipe from the tank, if there isn't one already installed. Fill the system and check for leaks before you heat the water and check again when the water is hot.

SEE ALSO > Draining the system 8, Connecting pipes 21–2, 25, Compression joints 22, PTFE tape 22, Building Regulations on electrical wiring 70, Side-entry heaters 81, Replacing an element 82, Wiring immersion heaters 82

Unvented cylinders

An unvented cylinder supplies mains-pressure hot water throughout the house. This is achieved by connecting the cylinder directly to the rising main. Most manufacturers recommend a 22mm (3/4in) incoming pipe, but a 15mm (1/2in) main at high pressure is normally adequate. An unvented cylinder can be heated directly, using immersion heaters; or indirectly, provided you are not using a solid-fuel boiler.

Unvented cylinder

There are no storage tanks, feed-and-expansion tanks, or open-vent pipes associated with unvented cylinders. Instead, a diaphragm inside a pressure vessel mounted on top of the cylinder flexes to accommodate expanding water. If the vessel fails, an expansion-relief valve protects the system by releasing water via a discharge pipe.

There are several other safety devices associated with unvented cylinders. A normal thermostat should keep the temperature of the water in the cylinder below 65°C (150°F). If it reaches 90°C (195°F), then a second thermostat will either switch off the immersion heater or shut off the water supply from the boiler. Finally, if it should get as hot as 95°C (205°F), a temperature-relief valve opens and discharges water outside.

Bylaws and regulations

The installation of an unvented hot-water cylinder needs to comply with both the Water Regulations and the Building Regulations. It has to include all the necessary safety devices and be installed by a competent fitter, such as those registered with the Institute of Plumbing, the Construction Industry Training Board, or the Association of Installers of Unvented Hot Water Systems (Scotland and Northern Ireland). Have the installation serviced regularly by a similarly qualified fitter, to make sure all the equipment remains in good working order.

You must notify your water company and local Building Control Office if you intend to fit an unvented hot-water cylinder.

Thermal-store cylinder

A thermal-store cylinder reverses the indirect principle. Water heated by a central-heating boiler passes through the cylinder and transfers heat, via a highly efficient coiled heat exchanger, to mains-fed water supplying hot taps and showers. An integral feed-and-expansion tank is normally built on top of the cylinder.

When the system is working at maximum capacity, the mains-fed water is delivered at such a high temperature that cold water must be added via a thermostatic mixing valve plumbed into the outlet supplying taps and showers. As the cylinder is exhausted, less cold water is added. The thermal-store system provides mains-pressure hot water throughout the house, dispenses with the need for a cold-water storage tank in the loft, and increases the efficiency of the boiler.

A valve is needed to prevent the heat from the cylinder 'thermo-siphoning' (gravity circulating) around the central-heating system. This can be a motorized valve or a simple mechanical gravity-check (non-return) valve that is opened by the force of the central-heating pump.

As with all open-vented systems, the feed-and-expansion tank determines the head of water, and radiators must be lower than the tank in order to be filled with water. When the tank is combined with the cylinder, it needs to be situated on the top floor of the house in order to provide central heating throughout the building. If that is impossible, install a tankless thermal-store cylinder in your chosen location and fit a conventional feed-and-expansion tank in the loft.

A thermal-store cylinder

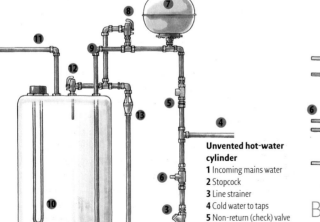

Unvented hot-water cylinder
1 Incoming mains water
2 Stopcock
3 Line strainer
4 Cold water to taps
5 Non-return (check) valve
6 Pressure limiter
7 Pressure vessel
8 Expansion-relief valve
9 Cold-water inlet
10 Immersion heater
11 Hot-water outlet
12 Temperature-relief valve
13 Air break (tundish)
14 Discharge pipe

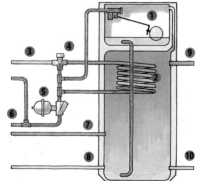

Thermal-store cylinder
1 Integral feed-and-expansion tank
2 Heat-exchanger
3 Supply pipe to hot taps/shower
4 Thermostatic mixing valve
5 Expansion vessel
6 Mains feed
7 Space-heating flow
8 Space-heating return
9 Boiler flow
10 Boiler return

Benefits of unvented systems

Thermal-store cylinders help to reduce boiler cycling and, because they heat water more quickly, can cut fuel usage by up to 15 per cent. Their capacities range from 80 to 210 litres (17½ to 46 gallons) and the systems can deliver flow rates of between 18 and 30L/min (31½ and 52½ pints). Unvented systems heat quickly and their high levels of insulation mean low heat losses. High flow rates in excess of 30L/min (52½ pints) are achievable.

SEE ALSO > Storage tanks 51, Wet central heating 55

Solar heating

Saving energy is a priority if we are to prevent further damage to our environment from the effects of carbon dioxide. The systems that have been developed to harness solar energy offer an effective alternative for heating domestic water. In contrast to the demand for space heating, which varies according to the season, hot water is required constantly throughout the year – and is well suited to heating with solar energy.

Using solar energy to heat water

The idea of using the sun to provide free, non-polluting energy for heating water has always appealed to conservationists, but thanks to soaring energy costs and technological developments, solar power is becoming increasingly popular.

There are two main types of solar panel: solar collectors, which absorb energy from the sun and use it to heat water; and photovoltaic or solar electric panels, which convert solar radiation directly into electricity. Solar collectors are of most use in the home and there are two types: flat plate and evacuated tubes.

Flat plate collectors can be a simple sheet of metal painted black that absorbs the sun's energy. Water is fed through the panel in pipes attached to the metal sheet (a central heating radiator is often used in DIY panels) and picks up the heat in the metal. The metal sheet is embedded in an insulated box and covered with glass or clear plastic on the front.

The evacuated tube system is more expensive and uses a series of glass heat tubes grouped together. The tubes are highly insulated, due to a vacuum inside the glass. From the late spring through to early autumn, this type of system can produce a useful proportion of the household hot water. During the winter,

Mount collectors on a south-facing roof

the solar collectors provide useful 'preheat' that reduces the time it takes a boiler to heat water, thereby saving energy.

There are a number of companies that supply solar collectors for heating water, plus all the controls and pipework required to complete the job. If you carry out the plumbing yourself, the payback on the investment will be that much greater.

Ideally, a solar panel should be mounted on a south-facing roof at a 30° angle to the horizontal and out of reach of shadows from trees, buildings or chimneys.

A basic system

Most systems for supplying domestic hot water will require solar collectors that cover about 4sq m (4sq yd) of roof space. In order to trap maximum heat from the sun, the collectors should be mounted on a south-facing pitched roof. Collectors can be fitted, with minimal structural alterations, to almost any building, and planning approval is rarely required.

The most common way of utilizing solar energy to boost an existing water-heating system is to feed the hot water from the collectors to a second heat exchanger fitted inside your hot-water cylinder. This usually means replacing the cylinder with a dual-coil model.

An alternative technique is to plumb in a second well-insulated cylinder, which will 'preheat' the water before it is passed on to the main storage cylinder. This may involve raising the cold-water storage tank in order to feed the new preheat cylinder.

Controls

A pump is needed to circulate the water from the collectors to the cylinder coil and back to the collectors. A programmable thermostat, which operates the pump, senses when the panels are hotter than the water in the cylinder.

Point-of-use water heaters

Electric point-of-use water heaters are often designed to fit inside a cupboard or vanity unit beneath a sink or basin. You can install one of these heaters yourself, provided that it has a capacity of less than 9 litres (16 pints). Follow the manufacturer's instructions precisely, and fit a pressure-limiting valve and a filter (both of these are supplied as a kit). Also, make sure that the safety vent pipe discharges hot water to a place outside where it won't endanger anyone.

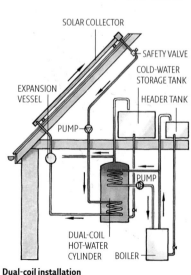

Dual-coil installation

SOLAR COLLECTOR
SAFETY VALVE
COLD-WATER STORAGE TANK
EXPANSION VESSEL
HEADER TANK
PUMP
PUMP
DUAL-COIL HOT-WATER CYLINDER
BOILER

Two-cylinder installation

COLD-WATER STORAGE TANK
PREHEAT CYLINDER
SOLAR COLLECTOR
SAFETY VALVE
HEADER TANK
PUMP
EXPANSION VESSEL
PUMP
BOILER
HOT-WATER CYLINDER

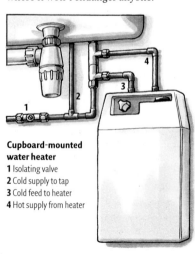

Cupboard-mounted water heater
1 Isolating valve
2 Cold supply to tap
3 Cold feed to heater
4 Hot supply from heater

SEE ALSO > Connecting pipes 21–2, 25, Hot-water cylinders 52, Building Regulations on electrical wiring 70, Wiring instantaneous heaters 79

Wet central heating

Wet central heating is economic, efficient and reliable and by far the most common form of central heating. It offers relatively quick warm-up times and a variety of control options for flexibility and energy saving. Although it is prone to various problems, most of these are easily resolved in a well-designed system.

Open-vented systems

The most common form of wet central heating is the two-pipe open-vented system – in which water is heated by a boiler and pumped through small-bore pipes to radiators or convector heaters, where the heat from the water is released into the rooms. The water then circulates back to the boiler for reheating, using natural gas, bottled gas (propane), oil, electricity, or a solid fuel.

Thermostats and valves allow the output of the individual heat emitters to be adjusted automatically, and parts of the system can be shut down as required.

This type of system can be used to heat the domestic hot-water supply, as well as the house itself. Some older systems employ gravity circulation to heat the hot-water storage cylinder but incorporate an electric pump to force the water around the radiators. In most modern systems, a pump propels the water to the cylinder and radiators via diverter valves.

Sealed systems

A sealed system is an alternative to the traditional open-vented method. Water is fed into the system via a filling loop, which is temporarily connected to the mains. In place of a feed-and-expansion tank (see top right), a pressure vessel containing a flexible diaphragm accommodates the expansion of the water as the temperature rises. Should the system become overpressurized, a safety valve discharges some of the water.

There is less corrosion in a sealed system and, because it runs at a relatively high temperature, the radiators can be smaller. Because there's no feed-and-expansion tank installed in the loft, radiators can be placed anywhere in the house, including in the loft itself.

However, sealed systems must be completely watertight – since there is no automatic top up – and they have to be made with expensive, high-quality components to prevent pressure loss. A boiler with a high-temperature cutout is required, in case the ordinary thermostat fails. Also, radiators get very hot.

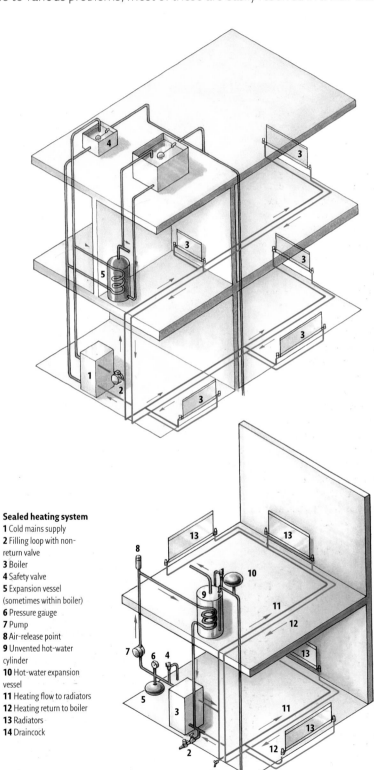

Sealed heating system
1 Cold mains supply
2 Filling loop with non-return valve
3 Boiler
4 Safety valve
5 Expansion vessel (sometimes within boiler)
6 Pressure gauge
7 Pump
8 Air-release point
9 Unvented hot-water cylinder
10 Hot-water expansion vessel
11 Heating flow to radiators
12 Heating return to boiler
13 Radiators
14 Draincock

Open-vented system
The water heated by the boiler (**1**) is driven by a pump (**2**) through a two-pipe system to the radiators (**3**) or special convector heaters, which give off heat as the hot water flows through them, gradually warming the rooms to the required temperature; the water then returns to the boiler to be reheated. A feed-and-expansion tank (**4**), situated in the loft, keeps the system topped up and takes the excess of water created by the system overheating. The hot-water cylinder (**5**) is heated by gravity circulation. In the diagram, red indicates the flow of water from the pump and blue shows the return flow.

● **One-pipe system**
In an outdated one-pipe system, heated water is pumped around the perimeter of the house through a single large-bore pipe that forms a loop. Flow and return pipes divert hot water to each radiator by means of gravity circulation. Larger radiators may be required at the end of the loop in order to compensate for heat loss. A one-pipe system incorporates a feed-and-expansion tank and a hot-water circuit similar to those used for conventional two-pipe systems.

Central-heating boilers

Technological improvements have made it possible to produce central-heating boilers that are smaller and more efficient than their predecessors. Today, gas and oil are still the most popular fuels because, despite advances in solid-fuel technology, the dirt and inconvenience associated with solid fuels can't be ignored or overcome. Wood-burning boilers aren't really practical for suburban homes, though a wood-fired stove could be used to heat a room, perhaps with a small back boiler to provide hot water.

Condensing boilers

Condensing boilers extract more heat from the fuel than other types of boiler. This is achieved either by passing the water through a highly efficient heat exchanger or by having a secondary heat exchanger that uses heat from the flue to 'preheat' cool water returning from the radiators.

With a conventional boiler, the moisture within the exhaust gases passes through the flue as steam. In a condensing boiler, which extracts additional heat from the gases, the moisture condenses within the boiler and is drained through a pipe.

Building Regulations now require all new gas- and oil-fired boilers fitted in households to be of the condensing type, because of their low energy use, whether the installation is a new build or a replacement. More complex than a standard gas boiler, they are also more costly.

Strict rules apply on the location of the flue to avoid gases being vented across footpaths or under a neighbour's window, for instance. Your Building Control Officer will be able to advise.

Oil-fired boilers

Pressure-jet oil-fired boilers are fitted with controls similar to the ones for gas boilers described above. Oil boilers can be floor-standing or wall-mounted. Oil-fired heating requires a large external storage tank, with easy access for delivery tankers.

Gas-fired boilers

Older gas-fired boilers have pilot lights that burn constantly, in order to ignite the burners whenever heat is required. The burners may be operated manually or by a timer set to switch the heating on and off at selected times. The boiler can be linked to a room thermostat, so that the heating is switched on and off to keep temperatures at the required level throughout the house. Another thermostat inside the boiler prevents overheating.

An increasing number of boilers have energy-saving electronic ignition. The pilot is not lit until the room thermostat demands heat and cuts off once up to temperature.

Solid-fuel boilers

Solid-fuel boilers are invariably floor-standing and require a conventional flue. Back boilers are small enough to be built into a fireplace. The rate at which the fuel is burnt is controlled by a thermostatic damper or sometimes by a fan.

The system must have some means for the heat to escape in case of a circulation pump failure, or the water could boil, causing damage. This is usually in the form of a pipe that leads from the boiler and the heat exchanger in the domestic hot-water cylinder to a radiator situated in the bathroom, where the excess heat can be used to dry wet towels.

Some models have a hopper feed that tops them up automatically. You need a suitable place to store fuel for the boiler and the residual ash has to be removed regularly.

Combination boilers

Combination boilers provide both hot water to a sealed heating system and a separate supply of instant hot water directly to taps and showers. The advantages are ease of installation (no tanks or pipes in the loft), space-saving (no hot-water storage cylinder), economy (you heat only the water you use) and mains pressure at taps and shower heads.

The main drawback is a fairly slow flow rate, so filling a bath can take a long time and it's not usually possible to use two hot taps at the same time. Combination boilers are therefore best suited to small households or flats. However, to overcome these problems, the newer generation of combination boilers incorporate a small hot-water storage tank.

• Gas installers

Gas boilers must be installed by competent fitters registered with CORGI (Council for Registered Gas Installers). Check, also, that your installer has the relevant public-liability insurance for working with gas.

• Boiler flues

All boilers need some means of expelling the combustion gases that result from burning fuel. Frequently this is effected by connecting the boiler to a conventional flue or chimney that takes the gases directly to the outside.

Alternatively, some boilers, known as room-sealed balanced-flue boilers, are mounted on an external wall and the flue gases are passed to the outside through a short horizontal duct. Balanced-flue ducts are divided into two passages – one for the outgoing flue gases, and the other for the incoming air needed for efficient combustion.

All boilers can be connected to a conventional flue, but gas and oil-fired boilers are also made for balanced-flue systems. If the boiler is fan-assisted, it can be mounted at some distance – typically 3 or 4m (10 or 12ft) – from the balanced-flue outlet.

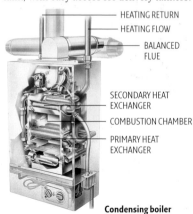

HEATING RETURN
HEATING FLOW
BALANCED FLUE
SECONDARY HEAT EXCHANGER
COMBUSTION CHAMBER
PRIMARY HEAT EXCHANGER

Condensing boiler

Heating requirements

The capacity (heat output) of the boiler needed can be calculated by adding up the specified heat output of all the radiators, plus a 3kW allowance for a hot-water cylinder. Ten per cent is added to allow for very cold weather. The overall calculation is affected by the heat lost through the walls and ceiling, and also by the number of air changes caused by ventilation.

Some plumbers' merchants will make the relevant calculations for you, if you provide them with the dimensions of each room. Alternatively, you can find calculators online.

The efficiency of a boiler is determined by its SEDBUK (Seasonal Efficiency of Domestic Boilers in the UK) rating from A to G. A-rated boilers are more than 90 per cent efficient.

SEE ALSO > Thermostats 59

Ventilating a boiler

A boiler that takes its combustion air from within the house and expels fumes through a conventional open flue (see opposite far left) must have access to a permanent ventilator fitted in an outside wall. The ventilator has to be of the correct size – as recommended by the boiler manufacturer – and must not contain a fly-screen mesh, which could become blocked. Refer to Building Regulations F1–1.8 for specific guidance. A boiler that is starved of air will create carbon monoxide – a lethal invisible gas that has no smell.

A cupboard that houses a balanced-flue room-sealed boiler must be fitted with ventilators at the top and bottom, to prevent the boiler overheating.

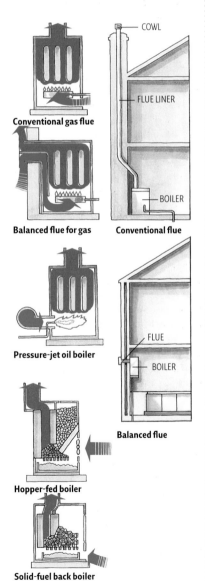

Conventional gas flue

Balanced flue for gas　　**Conventional flue**

COWL

FLUE LINER

BOILER

Pressure-jet oil boiler

FLUE

BOILER

Balanced flue

Hopper-fed boiler

Solid-fuel back boiler

Radiators

The hot water from a central-heating boiler is pumped along small-bore pipes connected to radiators, mounted at strategic points to heat individual rooms and hallways. The standard radiator is a double-skinned pressed-metal panel, which is heated by the hot water that flows through it.

Despite its name, a radiator emits only a fraction of its output as radiant heat – the rest being delivered by natural convection as the surrounding air comes into contact with the hot surfaces of the radiator. As the warmed air rises towards the ceiling, cooler air flows in around the radiator, and this air in turn is warmed and moves upwards. As a result, a very gentle circulation of air takes place in the room, and the temperature gradually rises to the optimum set on the room thermostat.

Panel radiators

Radiators are available in a wide range of sizes. The larger they are, the greater their heat output. Output for a given size can be increased further by using 'double radiators', which are made by joining two panels one behind the other. Most types of radiator have fins attached to their rear faces to induce convected heat.

The handwheel valve at one end of the radiator turns the flow of water on or off; the lockshield valve at the other end is set to balance the system, then left alone. An ordinary handwheel valve can be fitted at either end of a radiator, regardless of the direction of flow. However, thermostatic valves, which regulate the temperature of individual radiators, are marked with arrows to indicate the direction of flow and must be fitted accordingly.

A bleed valve is used to release air that can build up inside the radiator and prevent the panel from heating properly.

Decorative radiators
Radiators don't have to be white and flat. You could choose from modern takes on old-fashioned cast iron radiators, elegant vertical radiators in stainless steel, or curved, finned models designed for tight spaces or just to look good.

Double-panel radiator　　　　**Finned radiator**

Panel radiator
1 A manual handwheel valve turns the flow on or off.
2 A lockshield valve is set to balance the system.
3 A bleed valve disperses airlocks.

Heat emission
As it's heated by the radiator, convected air flows upwards and is replaced by cooler air near the base of the radiator. In addition, heat radiates from the surface of the panel.

Convectors and skirting radiators

Convectors and skirting radiators are a space-saving and more expensive alternative to panel radiators. Heat from hot water pumped through them is transferred to the air via fins or a panel, which replaces the skirting board in the room.

Convectors

Convector heaters can be used as part of a wet central-heating system. Some models are designed for inconspicuous fixing at skirting level.

Convectors emit none of their heat in the form of direct radiation. The hot water from the boiler passes through a finned pipe inside the heater, and the fins absorb the heat and transfer it to the air around them. The warmed air passes through a damper-controlled vent at the top of the heater, and at the same time cool air is drawn in through the open bottom to be warmed in turn.

With a fan-assisted convector heater, the airflow is accelerated over the fins in order to speed up room heating.

Rising warm air draws in cool air below

Fan-assisted convector
This unit makes good use of the wasted space under a kitchen unit.

Skirting radiators

A skirting radiator is a space-saving alternative to a conventional panel radiator and is designed for installation in place of a wooden skirting board. The twin copper-lined waterways and the outer casing are formed from a single aluminium extrusion (see below) and are clipped to the wall with a special bracket.

Made in various lengths up to 6m (19ft 6in) lengths and available in various finishes, skirting radiators are cut to length then joined at the corners of the room, using conventional soldered pipe joints. The pipework and valves are hidden from view, but are readily accessible for maintenance or repair.

Electric versions are also available.

Positioning convectors and radiators

At one time, central-heating radiators and convector units were nearly always placed under windows, because the area around a window tends to be the coldest part of a room. However, if you've fitted double glazing to reduce heat loss and draughts, then it will be more efficient to place heaters elsewhere – especially if windows are hung with long curtains.

Finned radiators offer a greater heat output and more flexibility in their siting. The shape of a room can also affect the siting of radiators and, perhaps, the number. For example, it is difficult to heat a large L-shaped room with just a single radiator at one end. In situations like this it's probably best to consult a heating installer to help you decide upon the optimum number and position of heaters.

Wherever possible, avoid hanging curtains or standing furniture in front of a radiator or convector heater. Curtains and furniture absorb radiated heat – and curtains also trap convected heat.

The warm air rising from a radiator will eventually discolour the paint or wall-covering above it. Fitting a narrow shelf about 50mm (2in) above a radiator avoids staining, without inhibiting convection. Alternatively, enclose the radiator in a decorative cover – heat output is barely reduced, provided air is able to pass through the enclosure freely, especially at the top and bottom (see below).

Radiator covers

Fitting a cover around a radiator will turn it into a decorative feature. The cabinet must be ventilated to allow air into the bottom and for the convected warm air to exit from the top. A perforated panel is usually fitted across the front to dissipate the heat and add to the unit's appearance.

Cabinets are available in kit form to fit standard-size radiators. Alternatively, you can make your own from MDF board.

Making your own cabinet
A radiator cabinet can be designed to stand on the floor or to be hung on the wall at skirting height. A floor-standing version is described here.

Cut the shelf (**1**) and two end panels (**2**) from 18mm (¾in) MDF. Make these components large enough to enclose the radiator and both valves.

Glue the panels to the shelf with dowel joints, and dowel a 50 x 25mm (2 x 1in) tie rail (**3**) between the sides at skirting level. Cut a new skirting moulding with a vent in the bottom edge (**4**) to fit along the base. Complete the box by applying a decorative moulding (**5**) around the edge of the shelf.

Cut a front panel (**6**) from either perforated hardboard, MDF, aluminium mesh or bamboo lattice, and mount it in a rebated MDF frame (**7**). Hold the frame in place with magnetic catches.

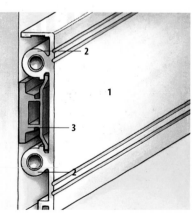

Skirting radiator
1 Aluminium extrusion
2 Copper-lined waterways
3 Mounting assembly

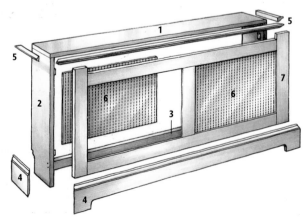

Floor-standing radiator cabinet
1 Shelf
2 End panel
3 Tie rail
4 Skirting
5 Moulding
6 Perforated panel
7 Frame

Controls for central heating

There are various automatic control systems and devices available for wet central heating that bring greater control and can help cut running costs by reducing wastage of heat to a minimum.

Automatic controllers can be divided into three basic types: temperature controllers (thermostats), automatic on-off switches (programmers and timers), and heating-circuit controllers (zone valves). These devices can be used, individually or in combination, to provide a very high level of control.

Thermostats

All boilers incorporate thermostats to prevent overheating. An oil-fired or gas boiler will have one that can be set to vary heat output by switching the unit on and off; and some models are also fitted with modulating burners, which adjust flame height to suit heating requirements. On a solid-fuel boiler, the thermostat opens and closes a damper that admits more or less air to the firebed to increase or reduce the rate of burning, as required.

A room thermostat – 'roomstat' for short – is often the only form of central-heating control fitted. It is placed in a room where the temperature usually remains fairly stable, and works on the assumption that any rise or drop in the temperature will be matched by similar variations throughout the house.

Roomstats control the temperature by means of simple on-off switching of the boiler – or the pump, if the boiler has to run constantly in order to provide hot water. The main drawback of a roomstat is that it can't sense temperature changes in other rooms - caused, for example, by the sun shining through a window or a separate heater being switched on.

More sophisticated temperature control is provided by a thermostatic valve, which can be fitted to a radiator instead of the standard manually operated valve. A temperature sensor opens and closes the valve, varying the heat output to maintain the desired temperature in the individual room. Thermostatic radiator valves need not be fitted in every room, but could reduce the heat in a kitchen or spare bedroom, for example, while a roomstat regulates the temperature in the rest of the house.

The most sophisticated thermostatic controller is a boiler-energy manager or 'optimizer', which collects data from sensors inside and outside the building in order to deduce the optimum running period so the boiler is not wastefully switched rapidly on and off.

Boiler-energy manager

Room thermostat

Programmer or timer

Thermostatic radiator valve

Heating controls
There are a number of ways to control heating:
1 A wiring centre connects the controls in the system.
2 A programmer/timer is used in conjunction with a zone valve to switch the boiler on or off at pre-set times, and run the heating and hot-water systems.
3 Optional boiler-energy manager controls the efficiency of the heating system.
4 Room thermostats are used to control the pump, or zone valves to regulate the overall temperature.
5 A non-electrical thermostatic radiator valve controls the temperature of an individual heater.

Zone-control valves

In most households the bedrooms are unoccupied for much of the day, so don't need heating continuously.

One way of avoiding such waste is to divide your central-heating system into circuits or 'zones' (upstairs and downstairs, for example) and to heat the whole house only when necessary. If you zone your house, make sure the unheated areas are adequately ventilated, to prevent condensation.

Control is via motorized valves linked to a timer or programmer that directs the heated water through selected pipes at predetermined times of the day. Alternatively, zone valves linked to individual thermostats can be used to provide separate temperature control for each zone.

A motorized zone-control valve

Timers and programmers

You can cut fuel bills substantially by ensuring that the heating is not on while you are out or asleep. A timer can be set so that the system is switched on to warm the house before you get up and goes off just before you leave for work, then comes on again shortly before you return home and goes off at bedtime. The simpler timers provide two 'on' and two 'off' settings, which are normally repeated every day. A manual override enables you to alter the times for weekends and other changes in routine.

More sophisticated devices, known as programmers, offer a larger number of on-off programs – even a different one for each day of the week – as well as control of domestic hot water.

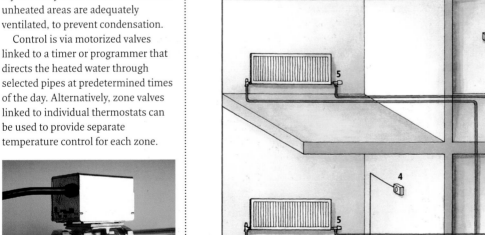

SEE ALSO > Radiators 57

Diagnosing heating problems

When heating systems fail to work properly, they can exhibit all sorts of symptoms, some of which can be difficult to diagnose without specialized knowledge and experience. However, it pays to check out the more common faults, summarized below, before calling out a heating engineer.

Hissing or banging sounds from boiler or heating pipes

This is caused by overheating due to:
• **Blocked chimney (if you have a solid-fuel boiler)** Sweep chimney to clear heavy soot.
• **Build-up of scale due to hard water** Shut down boiler and pump. Treat system with descaler. Drain, flush and refill system.
• **Faulty boiler thermostat** Shut down boiler. Leave pump working to circulate water, to cool system quickly. When it's cool, operate boiler thermostat control. If you don't hear a clicking sound, call in an engineer.
• **Lack of water in system** Shut down boiler. Check feed-and-expansion tank in loft. If empty, the valve may be stuck. Move float-valve arm up and down to restore flow and fill system. If this has no effect, check to see if mains water has been turned off by accident or (in winter) if supply pipe is frozen.
• **Pump not working (with a solid-fuel boiler)** Shut down boiler, then check that pump is switched on. If pump is not running, turn off power and check wired connections to it. If pump seems to be running but outlet pipe is cool, check for airlock by opening pump bleed screw. If pump is still not working, shut it down, drain system, remove pump and check for blockage. Clean pump or, if need be, replace it.

Radiators in one part of the house do not warm up

• **Timer or thermostat that controls relevant zone valve is not set properly or is faulty** Check timer or thermostat setting and reset if need be. If this has no effect, switch off power supply and check wired connections. If that makes no difference, replace unit.
• **Zone valve is faulty** Replace motor or drain system and replace entire valve.
• **Pump not working** *See above.*

All radiators remain cool, though boiler is operating normally

• **Pump not working** Check pump by listening or feeling for motor vibration. If pump is running, check for airlock by opening bleed valve. If this has no effect, the pump outlet may be blocked. Switch off boiler and pump, remove pump and clean or replace as necessary. If pump is not running, switch off and try to free spindle. Look for a large screw in the middle – removing or turning it will reveal the slotted end of the spindle. Turn this until the spindle feels free, then switch pump on again.
• **Pump thermostat or timer is set incorrectly or is faulty** Adjust thermostat or timer setting. If that has no effect, switch off power and check wiring connections. If they are in good order, call in an engineer.

Area at top of radiator stays cool

• **Airlock at top of radiator is preventing water circulating fully** Bleed radiator to release trapped air.

Single radiator doesn't warm up

• **Handwheel valve is closed** Open the valve.
• **Thermostatic radiator valve is set too low or is faulty** Adjust valve setting. If this has no effect, drain the system and replace the valve.
• **Lockshield valve not set properly** Remove lockshield cover and adjust valve setting until radiator seems as warm as those in other rooms. Have lockshield valve properly balanced when the system is next serviced.
• **Radiator valves blocked by corrosion** Close both radiator valves, remove radiator and flush out.

Cool patch in centre of radiator

• **Deposits of rust at bottom of radiator are restricting circulation of water** Close both radiator valves, remove radiator and flush out.

Boiler not working

• **Thermostat set too low** Check that roomstat and boiler thermostats are set correctly.
• **Programmer or timer not working** Check programmer or timer is switched on and set correctly. Replace if fault persists.
• **Gas boiler's pilot light goes out** Relight pilot following instructions supplied with boiler (these are usually printed on the back of the front panel). If pilot fails to ignite, have it replaced.

Continuous drip from overflow pipe of feed-and-expansion tank

• **Faulty float valve or leaking float, causing valve to stay open** Shut off mains water supply to feed-and-expansion tank and bale out to below level of float valve. Remove valve and fit new washer. Or, unscrew leaking float from arm and fit new one.
• **Leaking heat-exchanger coil in hot-water cylinder** Dripping from overflow will occur only if the feed-and-expansion tank is below the cold-water storage tank. Turn off boiler and mains water. Let system cool, then take dip-stick measurement in both tanks. Don't use water overnight – check again in morning. If water level has risen in the feed-and-expansion tank and dropped in the cold-water storage tank, replace hot water cylinder.

Water leaking from system

• **Loose pipe unions at joints, pump connections, boiler connections, etc.** Turn off boiler (or close down solid-fuel appliance) and switch off pump, then tighten leaking joints. If this has no effect, drain the system and remake joints.

SEE ALSO ➤ Curing leaks 9, Pipework 19–25, Plumbing joints 20–2, 24–5, Draining the system 61, Bleed valve 63, Removing a radiator 63, Replacing radiators 65

Draining and refilling

Although it's inadvisable to do so unnecessarily, there may be times when you have to drain your wet central-heating system completely and refill it. This could be for routine maintenance, when dealing with a fault, or because you have decided to extend the system or upgrade the boiler. The job can be done fairly easily if you follow the procedures outlined here.

Draining the system

Before draining your central-heating system, cool the water by shutting off the boiler and leaving the circulation pump running. The water in the system will cool quite quickly.

Switch off the pump and turn off the mains water supply to the feed-and-expansion tank in the loft either by closing the stopcock in the feed pipe or by laying a batten across the tank and tying the float arm to it.

The main draincock for the system will normally be in the return pipe near the boiler. Push one end of a garden hose onto its outlet and lead the other end of the hose to a gully or soakaway in the garden, then open the draincock. If you have no key for its square shank, use an adjustable spanner.

Most of the water will drain from the system, but some will be held in the radiators. To release the trapped water, start at the top of the house and carefully open the radiator bleed valves. Air will flow into the tops of the radiators, breaking the vacuum, and the water will drain out. Last of all, drain inverted pipe loops (see below).

Refilling the system

Before refilling the system, check all the draincocks and bleed valves are closed.

Restore the water supply to the feed-and-expansion tank in the loft. Filling the system will trap air in the tops of the radiators – so when full, bleed all the radiators, starting at the bottom of the house. You may also have to bleed the circulating pump. Finally, check all the draincocks and bleed valves for leaks, and tighten them if necessary.

Draincock key
A special tool is available for operating draincocks.

Draining procedure
Turn off the mains supply to the tank at the feed-pipe stopcock (**1**). If there's no stopcock, tie the float-valve arm to a batten laid across the tank (**2**). With a hose pushed onto the main draincock (**3**) and its other end at a gully or soakaway outside, open the draincock and let the system empty. Release any water trapped in the radiators (**4**) by opening their bleed valves (**5**), starting at the top of the house. Be sure to close all draincocks before you refill the system.

Cleaning the system

After installing or modifying a central-heating system, flush the pipework to get rid of swarf and flux, which can induce corrosion and damage valves or the pump.

Turn the pump impeller with a screwdriver before running the system in order to make sure it's clear. If you can feel resistance, drain the system and remove the pump before cleaning and refitting.

Descaling

If your system is old or badly corroded, a harsh cleaner or descaler may expose minor leaks sealed by corrosion – so use a mild cleanser, introduced into the system via the feed-and-expansion tank or inject it into a radiator via the bleed valve. Manufacturers' instructions vary, but in general the cleanser is left in the system for a week, with the boiler set to a fairly high temperature.

Afterwards, turn off and drain the system, then refill and drain it several times – if possible, using a hose to run mains-pressure water through the system while draining it. Some cleansers must be neutralized before you can add a corrosion inhibitor.

If your boiler is making loud banging noises, treat it with a fairly powerful descaler, running the hot-water programme only.

● **Power-flushing the system**
After upgrading an older system, perhaps with a new boiler or radiators, you could flush the system yourself (see left), but it's advisable to have it cleansed thoroughly by a heating engineer, using a power-flushing unit. When it is connected, the unit pumps chemically-treated water through the system to flush out impurities.

Inverted pipe loops

Often when fitting a central-heating system in a house that has a solid ground floor, installers run the heating pipes from the boiler into the ceiling void and drop them down the walls to the individual radiators. These 'inverted pipe loops' have their own draincock and must be drained separately after the main system has been emptied.

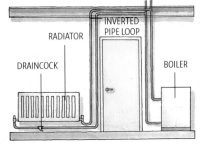

RADIATOR

INVERTED PIPE LOOP

DRAINCOCK

BOILER

An inverted pipe loop has its own draincock

SEE ALSO > Gully 7, Turning off the water 8, Bleeding radiators 63, Bleeding a pump 66

Maintaining your boiler

Oil-fired and gas boilers must be serviced annually to work efficiently and this work must be carried out by a qualified engineer. With either type of boiler, you can enter into a contract for regular maintenance with your fuel supplier or the original installer.

Corrosion in the system

- **Servicing gas boilers**
Any maintenance that involves dismantling any part of a gas boiler must be carried out by a CORGI-registered engineer, who should undertake all the necessary gas-safety checks as part of the service. There's no point in attempting to service the boiler yourself if you are not qualified and equipped to do so – it can also be dangerous, and you will be breaking the law.

Modern boilers and radiators are made from fairly thin materials, and if you fail to take basic anti-corrosion measures, the life of the system can be reduced to 10 years or less. Corrosion may result either from hard-water deposits or from a chemical reaction between the water and the system's metal components.

Lime scale
Scale builds up quickly in hard-water areas of the country. Even a thin layer of lime scale on the inner wall of a boiler's heat exchanger reduces its efficiency and may cause banging and knocking within the system. In fact, the scale can insulate sections of the heat exchanger to such an extent that it produces 'hot spots', leading to premature failure of the component.

Rust
Rust corrodes steel components, most notably radiators. Most rusting occurs within weeks of filling the system; but if air is being sucked in constantly, then rusting is progressive. Having to bleed radiators regularly is a sure sign that air is being drawn into the system.

Sludge
Magnetite (black sludge) clogs the pump and builds up in the bottom of radiators, reducing their heat output.

Electrolytic action
Dissimilar metals, such as copper and aluminium, act like a battery in the acidic water that is present in some central-heating systems. This results in corrosion.

Reducing corrosion

Drain about 600ml (1 pint) of water from the boiler or a radiator. Orange water denotes rusting, and black the presence of sludge. In either case, treat immediately with corrosion inhibitor (see below).

If there are no obvious signs of corrosion, test for inhibitor levels by dropping two plain steel nails into a jar containing some water from the system, and place two similar nails in a jar of clean tap water. After a few days the nails in the tap water should rust; but if the system contains sufficient corrosion inhibitor, the nails in the system water will remain bright – if they don't, your system needs topping up with inhibitor. It is important to use the same product that is already present in the system – if you don't know what that is, drain and flush the system, then refill with fresh water and inhibitor.

Adding corrosion inhibitor
You can slow down corrosion by adding a proprietary corrosion inhibitor to the water. This is best done when the system is first installed – but the inhibitor can be introduced into the system at any time,

provided the boiler is descaled before doing so. If the system has been running for some time, it is better to flush it out first by draining and refilling it repeatedly until the water runs clean. Otherwise, drain off about 20 litres (4 gallons) of water – enough to empty the feed-and-expansion tank and a small amount of pipework – then pour the inhibitor into the tank and restore the water supply, which will carry the inhibitor into the pipes. About 5 litres (1 gallon) will be enough for most systems, but check the manufacturer's instructions. Finally, switch on the pump to distribute the inhibitor throughout the system.

Reducing scale
You can buy low-voltage coils to create a magnetic field that will prevent the heat exchanger of your boiler becoming coated with scale. However, unless you have soft water in your area, the only way to actually avoid hard water in the system is to install a water softener.

Before fitting any device to reduce scale, it is essential to seek the boiler manufacturer's advice.

Servicing schemes

It pays to have your central-heating system serviced regularly. Check the Yellow Pages for a suitable engineer, or ask the original installer of the system if they are willing to undertake the necessary servicing.

Gas installations
Gas suppliers offer a choice of servicing schemes for boilers, primarily provided to cover their own installations, but they will also service systems put in by other installers following a satisfactory inspection of the installation.

The simplest of the schemes provides for an annual check and adjustment of the boiler. If any repairs are found to be necessary, either at the time of the regular check or at other times during the year, then the labour and necessary parts are charged separately. But for an extra fee it is possible to have both free labour and free parts for boiler repairs at any time of year. The gas supplier will also extend the arrangement to include inspection of the whole heating system when the boiler is being checked, plus free parts and labour for repairs.

You may find that your installer or a local firm of CORGI heating engineers offers a similar choice of servicing and maintenance contracts.

Oil-fired installations
Installers of oil-fired central-heating systems and suppliers of fuel oil offer servicing and maintenance contracts similar to those outlined above for gas-fired systems. The choice of schemes available ranges from an annual check-up to complete cover for parts and labour whenever repairs are necessary.

As with the schemes for gas, it pays to shop around and make a comparison of the various services on offer and the charges that apply.

Solid-fuel systems
If you have a solid-fuel system, it is important to have the chimney and the flueway swept twice a year. The job is very similar to sweeping an open-fire chimney, access being either through the front of a room heater that has a back boiler or through a soot door in the flue pipe or chimney breast.

Once the chimney has been swept, clean out the boiler with a stiff brush, remove the dust and soot, lift out any broken fire bars and drop new ones in.

Locating gas boilers
Modern boilers fit snugly into standard kitchen cupboards.

SEE ALSO > Water softeners 50, Flushing the system 61

System repairs and maintenance

Trapped air prevents radiators from heating up fully, and a regular intake of air can cause corrosion. If a radiator feels cooler at the top than at the bottom, it's likely that a pocket of air has formed inside it and is impeding full circulation of the water. Getting the air out of a radiator – 'bleeding' it – is a simple procedure.

Bleeding a radiator

First switch off the circulation pump, otherwise it may suck air into the radiator when a bleed valve is opened.

Each radiator has a bleed valve at one of its top corners, identifiable by a square-section shank in the centre of the round blanking plug. If you don't have a key, they are readily available from any DIY shop or ironmongers. Some valves are simply opened with a screwdriver.

Use the key to turn the valve's shank anticlockwise about a quarter of a turn. It shouldn't be necessary to turn it further – but have a small container handy to catch spurting water, in case you open the valve too far. You will probably also need a rag to mop up water that dribbles from the valve and it's wise to protect carpets in case of a spill. Don't try to speed up the process by opening the valve further than necessary to let the air out – that is likely to produce a deluge of water.

You will hear a hissing sound as the air escapes. Keep the key on the shank of the valve and when the hissing stops and the first dribble of water appears, close the valve tightly.

Fitting an air separator

If you are having to bleed a radiator or radiators frequently, a large quantity of air is entering the system, which can lead to serious corrosion.

Check that the feed-and-expansion tank in the loft is not acting like a radiator and warming up when you run the central heating or hot water. This indicates that hot water is being pumped through the vent pipe into the tank and taking air with it back into the system. Fit an air separator in the vent pipe, linked to the cold feed from the feed-and-expansion tank.

If the pump is fitted on the boiler return pipe, it may be sucking in air through the unions or through leaking valve spindles.

Heating system with air separator
1 Cold-water storage tank
2 Feed-and-expansion tank
3 Air separator
4 Pump
5 Motorized valve
6 Hot-water cylinder
7 Boiler
8 Radiator flow
9 Radiator return

Blocked bleed valve

If no water or air comes out when you attempt to bleed a radiator, check whether the feed-and-expansion tank in the loft is empty. If the tank is full of water, then the bleed valve is probably blocked with paint.

Close the inlet and outlet valve at each end of the radiator, then remove the screw from the centre of the bleed valve. Clear the hole with a piece of wire, and reopen one of the radiator valves slightly to eject some water from the hole. Close the radiator valve again and refit the screw in the bleed valve. Open both radiator valves and test the bleed valve again.

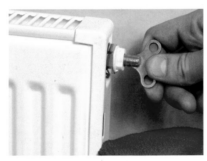

Dispersing an air pocket in a radiator

Removing a radiator

There are a number of reasons why it may be necessary to remove a radiator – for example, to make decorating the wall behind it easier. You can remove individual radiators without having to drain the whole system.

Make sure you have plenty of rag to hand for mopping up spilled water, plus a jug and a large bowl. The water in the radiator may be very dirty – so roll back the floorcovering before you start.

Shut off both valves, turning the shank of the lockshield valve clockwise with a key or an adjustable spanner (**1**). Note the number of turns needed to close it, so that later you can reopen it by the same amount.

Unscrew the cap-nut that keeps the handwheel valve or lockshield valve attached to the adaptor in the end of the radiator (**2**). Hold the jug under the joint and open the bleed valve slowly to let the water drain out. Transfer the water from the jug to the bowl, and continue until no more water can be drained off.

Unscrew the cap-nut that keeps the other valve attached to the radiator, lift the radiator free from its wall brackets, and drain any remaining water into the bowl (**3**).

After replacing the radiator, tighten the cap-nuts on both valves. Close the bleed valve and reopen both valves (open the lockshield valve by the same number of turns you used when closing it). Last of all, bleed the air from the radiator.

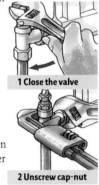

1 Close the valve

2 Unscrew cap-nut

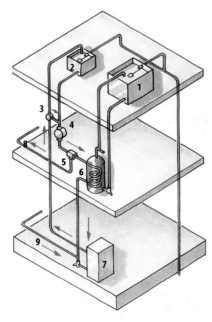

3 Lift radiator from brackets and drain

SEE ALSO > Draining the system 61, Filling the system 61

Replacing radiator valves

Like taps, radiator valves can develop leaks, although they're usually relatively easy to cure. Occasionally, however, it's necessary to replace a faulty valve.

Curing a leaking radiator valve

VALVE HEAD

SPINDLE

GLAND NUT

Leaking spindle
To stop a leak from a radiator-valve spindle, tighten the gland nut with a spanner. If the leak persists, undo the nut and wind a few turns of PTFE tape down into the spindle.

Water leaking from a radiator valve is probably seeping from around the spindle (see left). However, when the water runs round and drips from the valve's cap-nut, it's the nut that often appears to be the source of the leak. Dry the valve, then hold a paper tissue against the various parts of the valve to ascertain exactly where the moisture is coming from. If the nut is leaking, tighten it gently; if that's unsuccessful, undo and reseal it (see right).

Grip leaky valve with wrench and tighten cap-nut

Replacing a worn or damaged valve

To replace a lockshield or thermostatic valve, drain the system, then lay rags under the valve to catch the dregs. Holding the body of the valve with a wrench (or water-pump pliers), use an adjustable spanner to unscrew the cap-nuts that hold the valve to the pipe (**1**) and also to the adaptor in the end of the radiator. Lift the valve from the end of the pipe (**2**); if you're replacing a lock-shield valve, close it first – counting the turns, so you can open the new valve by the same number to balance the radiator.

Unscrew the valve adaptor from the radiator (**3**). You may be able to use an adjustable spanner, depending on the type of adaptor, or you may need a hexagonal radiator spanner.

Fitting the new valve
Ensure that the threads in the end of the radiator are clean. Drag the teeth of a hacksaw across the threads of the new adaptor to roughen them slightly, then wind PTFE tape four or five times around them. Screw the adaptor into the end of the radiator and tighten with a spanner.

Slide the valve cap-nut and a new olive over the end of the pipe and fit the valve (**4**) – but don't tighten the cap-nut yet. First, holding the valve body with a wrench, align it with the adaptor and tighten the cap-nut that holds them together (**5**). Then tighten the cap-nut that holds the valve to the water pipe and replace the valve head (**6**). Refill the system and check for leaks.

• **Resealing a cap-nut**
Drain the system and undo the leaking nut. Smear the olive with silicone sealant and retighten the cap-nut. Don't overtighten the nut or you may damage the olive. As an alternative to sealant, wind two turns of PTFE tape around the olive (not around the threads).

Replacing O-rings in a Belmont valve

The spindle of a Belmont valve is sealed with O-rings – which you can replace without having to drain the radiator. To find out which O-rings you need, take the plastic head of the valve to a plumbers' merchant before you begin work. On very old valves the rings are green, whereas the newer rings are red.

Undo the spindle (which has a left-hand thread). A small amount of water will leak out – but as you continue to remove the spindle, water pressure will seal the valve.

Prise off the O-rings, using a small screwdriver, and then lubricate the spindle with a smear of silicone grease. Slide on the new rings and replace the spindle.

Replacing O-rings
O-rings are housed in grooves in the valve spindle. Water pressure will seal the valve as the spindle is removed.

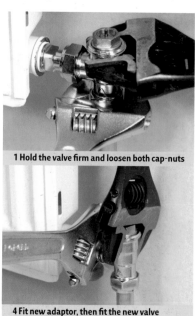
1 Hold the valve firm and loosen both cap-nuts

2 Unscrew the cap-nuts and lift the valve out

3 Remove the valve adaptor from the radiator

4 Fit new adaptor, then fit the new valve

5 Connect valve to adaptor and tighten cap-nut

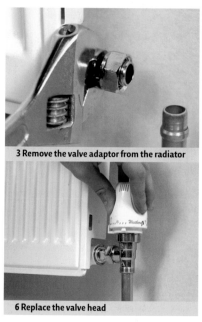
6 Replace the valve head

SEE ALSO > Draining the system 61

Replacing a radiator

Try to obtain a new radiator that is exactly the same size as the one you're planning to replace. This makes the job relatively easy and less disruptive, especially as the pipework may not have to be altered.

Simple replacement

Drain the old radiator and remove it from the wall. Unscrew the two valve adaptors at the bottom of the radiator, using an adjustable spanner or a hexagonal radiator spanner. Next, use a bleed key to unscrew the bleed valve; then remove both of the blanking plugs from the top of the radiator, using a radiator spanner (**1**).

Clean any corrosion or old PTFE tape from the threads of the adaptors and blanking plugs with wire wool (**2**), then wind four or five turns of new PTFE tape around the threads (**3**). Screw the plugs and adaptors into the new radiator, and then screw the bleed valve into its blanking plug.

Hang the new radiator on the wall brackets and connect the valves to their adaptors. Open the valves, then fill and bleed the radiator.

Installing a different-pattern radiator

More work is involved in replacing a radiator if you can't get one to match the existing one. You may have to fit new wall brackets and alter the pipe runs.

Drain your central-heating system, then take the old brackets off the wall. Lay the new radiator face down on the floor and slide one of its brackets onto the hangers welded to the back of the radiator. Measure the position of the brackets (**1**) and transfer these measurements to the wall. You need to allow a clearance of 100 to 125mm (4 to 5in) below the radiator.

Line up the new radiator brackets with the pencil marks on the wall, mark the fixing-screw holes for them, drill and plug the holes, then screw on brackets (**2**).

Take up the floorboards below the radiator and sever the vertical portions of the feed and return pipes (either cap the old T-joints or replace them with straight joints). Connect the valves to the bottom of the radiator and hang it on its brackets.

Slip a new vertical pipe into each of the valves and, using either capillary or compression fittings, connect these pipes to the original pipework running under the floor (**3**). Tighten the nuts connecting the new pipes to the valves.

Finally, refill the system with water, and check all the new connections and joints for leaks.

1 Measure the positions of the radiator brackets and transfer these dimensions to the wall

1 Remove the plugs, using a radiator spanner

2 Remove old PTFE tape and clean the threads

3 Tape the threads to make them watertight

2 Screw the mounting brackets to the wall

3 Align vertical section of pipe with radiator valve

SEE ALSO > Connecting pipes 20–2, 24–5, Draining the system 61, Bleeding radiators 63, Removing radiators 63

Servicing a pump

Wet central heating depends on a steady cycle of hot water being pumped from the boiler to the radiators and then back to the boiler for reheating. A faulty pump may result in poor circulation or none at all. Adjusting or bleeding the pump may be the answer; otherwise, it may need replacing.

Bleeding the pump

If an airlock forms in the circulation pump, the impeller will spin ineffectually and the radiators and hot-water tank will fail to warm up properly. The cure is to bleed the air from the pump, a procedure similar to bleeding a radiator. Have a jar handy to catch any spilled water and watch that none can spill on any electrical devices.

Switch off the pump and look for a screw-in bleed valve in the pump's outer casing. It may be located behind a cover. Open the bleed valve slightly with a screwdriver or vent key until you hear air hissing out. When the hissing stops and a drop of water appears, close the bleed valve.

Open the bleed valve with a screwdriver

Using a radiator thermometer
A pair of thermometers are used to measure the temperature drop across a radiator – one clipped to the feed and one to the supply pipes.

Adjusting the pump

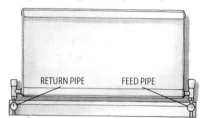

RETURN PIPE   FEED PIPE

Clip thermometers to the radiator pipes

Adjust pump speed to alter the temperature

There are two basic types of central-heating pump: fixed-head and variable-head. Fixed-head pumps run at a single speed, while the speed of variable-head pumps is adjustable.

When fitting a variable-head pump, the installer balances the radiators, then adjusts the pump's speed to achieve an optimum temperature for every room. If you can't boost a room's temperature by opening the radiator's handwheel valve, try adjusting the pump speed. However, before adjusting the pump, you should check that all your radiators show the same temperature drop between their inlets and outlets. To test your radiators, you can obtain a pair of clip-on thermometers from a plumbers'

merchant. Clip one of the thermometers to the feed pipe just below the radiator valve, and the other one to the return pipe, also below its valve. The difference between the temperatures registered by the thermometers should be about 11°C (20°F). If it's not, close the lockshield valve slightly to increase the difference in temperature; or open the valve to reduce it.

Having balanced all the radiators, you can now adjust the pump's speed by one increment at a time until the radiators are giving the overall temperatures you require. Depending on the make and model of pump, you may need to use a special tool, such as an Allen key, to make the adjustments. Switch off the pump before making each adjustment.

Replacing a worn pump

If you have to replace a faulty pump, make sure the new one is of a similar specification. A good plumbers' merchant will be able to advise.

First, turn off the boiler and close the isolating valves situated on each side of the pump. If the pump lacks isolating valves, you will have to drain down the whole system.

At your consumer unit, identify the electrical circuit that supplies the pump and remove the relevant circuit fuse or MCB. Then take the coverplate off the pump (**1**) and disconnect its wiring.

With a bowl or bucket ready to catch the water from the pump, undo the nuts that hold the pump to the valves or pipework (**2**).

● **Bridging the gap**
Modern pumps are sometimes smaller than equivalent older models. If this proves to be the case, buy a converter designed to bridge the gap in the existing pipework.

Having removed the old pump, install the new one (**3**), taking care to fit correctly any sealing washers that are provided. Tighten the connecting nuts.

Remove the coverplate from the new pump and feed in the flex. Connect the wires to the pump's terminals (**4**), then replace the coverplate. If the pump is of the variable-head type (see above), set the speed control to match the speed indicated on the old pump.

Open both isolating valves – or refill the system, if you had to drain it – then check the pump connections for leaks.

Open the pump's bleed valve to release any trapped air. Finally, replace the fuse or MCB in the consumer unit and test the pump.

1 Remove coverplate

2 Undo connecting nuts

3 Attach new pump

4 Connect power flex

SEE ALSO > Building Regulations on electrical wiring 70, Switching off the power 76, Draining the system 61, Filling the system 61

Replacing a control valve

Control valves add flexibility and control to central heating systems, but if they're worn or faulty, they can seriously impair the reliability of the system, and should be repaired or replaced promptly.

Replacing a faulty valve

Two-port control valve
A two-port valve seals off a section of pipework when the water has reached the required temperature.

Three-port control valve
This type of valve can isolate the central heating from the hot-water circuit.

When you buy a new valve, to avoid modifying pipework, make sure it is of the same pattern as the one you are replacing.

Drain the system. Then, at your consumer unit, remove the fuse or MCB for the circuit to which the central-heating controls are connected.

The flex from the valve will be wired to an adjacent junction box, which is also connected to the heating system's other controls. Take the cover off the box and disconnect the wiring for the valve – making a note of the terminals used, to make reconnection easier.

To remove the old valve, simply cut through the pipe on each side (**1**). When fitting the new valve, bridge the gap with short sections of pipe, complete with joints at each end (**2**). Spring the assembly into place and connect the joints to the old pipe, then tighten the valve cap-nuts (**3**). Connect the valve's flex to the junction box, then insert the circuit fuse or MCB.

1 If you're unable to disconnect the valve, use a hacksaw to cut through the pipe on each side

2 With the new valve connected to short sections of pipe, spring the assembly into the pipe run

3 Having connected the pipes, tighten the valve cap-nuts on each side, using a pair of adjustable spanners

Replacing the electric motor

If a motorized valve ceases to open, its electric motor may have failed. Before replacing the motor, use a mains tester to check whether it's receiving power. If it is, fit a new motor.

There is no need to drain the system. Switch off the electricity supply to the central-heating system – don't merely turn off the programmer, as motorized valves have a permanent live feed.

Once the power is off, remove the cover and undo the single screw that secures the motor (**1**). Open the valve, using the manual lever, and lift out the motor (**2**). Disconnect the two motor wires by cutting off the connectors.

Insert a new motor (available from a plumbers' merchant), then let the lever spring back to the closed position. Refit the retaining screw. Strip the ends and connect the wires, using the new connectors supplied (**3**). Replace the cover, and test the operation by turning on the power and running the system.

Compression coupling

Soldered coupling

Slip couplings
It can sometimes be difficult to replace a valve using two conventional joints. If you can't spring the new assembly into place (see far left), use a slip coupling at one end. This coupling is free to slide along the pipe to bridge the gap.

1 Remove cover and then retaining screw   **2 Push lever to open valve; lift out motor**

3 Join the wires, using the two connectors supplied

SEE ALSO > Building Regulations on electrical wiring 70, Switching off the power 76, Making pipe joints 20–2, 24–5, Heating controls 59, Draining the system 61

Underfloor heating

Thanks to flexible plastic plumbing, sophisticated controls and efficient insulation, underfloor heating is becoming a more popular and affordable form of central heating. Most systems are intalled as the house is being built, but specialist manufacturers have developed a range of warm-water heating systems to suit virtually any situation.

Underfloor-heating systems

Underfloor heating can be incorporated into any type of floor construction, including solid-concrete floors, boarded floating floors and suspended timber floors (see below). The heat emanates from a continuous length of plastic pipe that snakes across the floor – like an elongated radiator – across one or more rooms.

The entire floor area is divided into separate zones to provide the most efficient layout. Each zone is controlled by a roomstat and is connected to a thermostatically controlled multi-valve manifold that forms the heart of the system. The manifold controls the temperature of the water and the flow rate to the various zones. Once a room or zone reaches its required temperature, a valve automatically shuts off that part of the circuit. A flow meter for each of the zones allows the circuits to be balanced when setting up the system and subsequently monitors its performance.

The manifold, which is installed above floor level, is connected to the boiler via a conventional circulation pump.

• Combining systems
You can have radiators upstairs and underfloor heating downstairs. A mixing manifold will allow you to combine the two systems, using the same boiler. Any type of boiler is suitable for underfloor heating, but a condensing boiler is the most economical.

Benefits of underfloor heating

Although it's easier to incorporate underfloor heating while a house is being built, installing it in an existing building is possible, though you may have to raise floors to accommodate it. Underfloor heating can be made to work alongside a panel-radiator system and can provide the ideal solution for heating a new conservatory, for example.

Compared with panel radiators, an underfloor-heating system radiates heat more evenly and over a wider area. This has the effect of reducing hot and cold spots within the room and produces a more comfortable environment, where the air is warmest at floor level and cools as it rises towards the ceiling.

Underfloor heating is more energy-efficient and less costly to run than other central-heating systems because it operates at a lower temperature – and because there's a more even temperature throughout a room, the roomstat can be set a degree or two lower. With relatively cool water in the return cycle, a modern condensing boiler works even more efficiently.

Methods for installing underfloor heating

When underfloor heating is installed in a new building, the plastic tubes are usually set into a solid-concrete floor. Flooring insulation is laid over the base concrete, and rows of special pipe clips are fixed to the insulation; sometimes a metal mesh is used instead of the clips. The flexible heating tubes are then clipped into place at the required spacings (see opposite), and a concrete screed is poured on top.

With a boarded floating floor, a layer of grooved insulation is laid over the concrete base, and the pipes are set in aluminium 'diffusion' plates inserted in the grooves. The entire floor area is then covered with an edge-bonded chipboard or a similar decking material.

The heating pipes can be fastened with spacer clips to the underside of a suspended wooden floor. In this situation, clearance holes are drilled through the joists at strategic points to permit a continuous run of pipework. Reflective foil and thick blanket insulation is installed below the pipes.

It is possible to lay the pipes on top of a suspended floor, but this method raises the floor level by the thickness of the pipe assembly and the new decking.

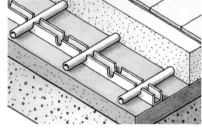

Screeded concrete floor

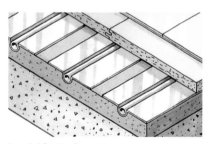

Boarded floating floor

Suspended wooden floor

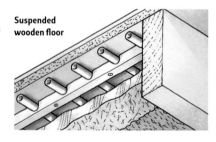

SEE ALSO > Panel radiators 57, Thermostats 59

Installing underfloor heating

Underfloor heating is a good choice for heating a new conservatory. The large areas of glass in a conservatory present very few options for placing radiators, and the concrete slab that is typically used for conservatory floors provides an ideal base for this form of heating.

The basic plumbing system

Your supplier will suggest the best point to connect your new plumbing to the existing central-heating circuit. It can be at any convenient point, provided that the performance of your radiators will not be affected.

The pipework connecting the manifold for the underfloor heating to the radiator circuit can be metal or plastic, and it can be the same size as, but not larger than, the existing pipes. Again, your supplier will advise what to use.

The flow and return pipes from the manifold to the conservatory circuit (illustrated here, as an example) are connected to individual zone distributors, which in turn are connected to the flexible underfloor-heating tubes.

Basic system
1 Flow and return pipes from existing central-heating circuit.
2 Water-temperature mixing valve.
3 Pump
4 Manifold with zone valves.
5 Zone distributors.
6 Underfloor-heating tube.

Where to start

Send details of your proposed extension to the underfloor-heating supplier. In order to be able to supply you with a well-planned scheme and quotation, the company will also need a scaled plan of your house and the basic details of your present central-heating system.

Your options

The simplest system will connect to the pipework of your existing radiator circuit. Heat for the extension will be available only when the existing central heating is running, although the temperature in the conservatory can be controlled independently by a roomstat connected to a motorized zone valve and the underfloor-heating pump.

For full control, the flow and return pipework to the underfloor system must be connected directly to the boiler, and the roomstat must be wired up to switch the boiler on and off and to control the temperature of the conservatory.

Floor construction
1 Blinded hardcore
2 DPM
3 Concrete base
4 Insulation
5 Edge insulation
6 Pipe clips and pipe
7 Screed
8 Floor tiles

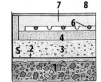

Constructing the floor

You will need to excavate the site and lay a concrete base as recommended by the conservatory manufacturer, a surveyor, or your local Building Control Officer. The base must include a damp-proof membrane.

Allow for a covering layer of floor insulation – a minimum of 50mm (2in) flooring-grade expanded polystyrene or 30mm (1¼ in) extruded polyurethane (check with your Building Control Officer). The floor should be finished with a 65mm (2½ in) sand-and-cement screed, plus the preferred floorcovering.

When laying the floor insulation slabs, you should install a strip of insulation, 25mm (1in) thick, all round the edges. This is to prevent cold bridging between the masonry walls and the floor screed.

Cut a hole through the house wall, ready for the new plumbing that will connect to the existing central-heating system.

Installing the system

Mount the manifold in a convenient place and connect the two distributor blocks below it – one for the flow, and the other for the return. Run the flow and return pipes back into the house, ready for connecting to the existing central-heating circuit. Install your new pump and a mixing valve in the flow and return pipes.

Following the layout supplied by the system's manufacturer, press the spikes of the pipe clips into the insulation at the prescribed spacing (**1**). Lay out the heating tubes for both coils, and clip them into place. Push the end of one of the coils into the flow distributor, and the other end of the same coil into the return distributor (**2**). Connect the other coil similarly.

Connect the flow and return pipes to the house's central-heating system – it pays to insert a pair of isolating valves at this point, so that you can shut off the new circuit for servicing. Fill, flush out and check the new system for leaks.

Apply the screed composed of 4 parts sharp sand : 1 part cement, with a plasticizer additive. Leave it to dry naturally for at least three weeks before laying your floorcovering.

Fit the roomstat at head height, out of direct sunlight, then make the electrical connections according to the instructions.

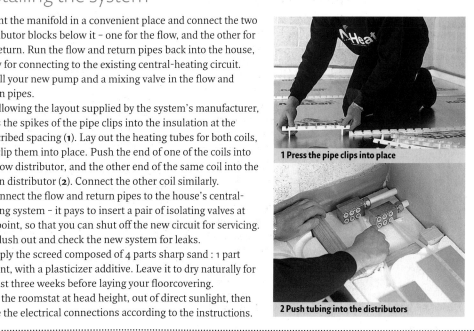

1 Press the pipe clips into place

2 Push tubing into the distributors

SEE ALSO > Building Regulations on electrical wiring 70, Switching off the power 76, Pipe joints 20–2, 24–5, Heating controls 59, Draining the system 61

First things first

Plumbing in kitchens, bathrooms and utility rooms often involves connecting appliances to the electricity supply. There are strict limitations on the type of electrical work you can undertake in these locations, and you should read the information on Building Regulations, below, before proceeding.

Understanding the basics

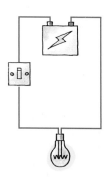

A basic circuit
Electricity runs from the source (battery) to the appliance (bulb) and then returns to the source. A switch breaks the circuit to interrupt the flow of electricity.

Electrical circuits are based on simple principles. For any electrical appliance to work, the power must be able to flow along a wire from its source to the appliance (say, a light bulb) and then back to the source along another wire. If the circuit is broken at any point, the appliance will stop working – the bulb will go out. Breaking the circuit – and restoring it as required – is what a switch is for. When the switch is in the 'on' position, the circuit is complete and the bulb operates. Turning the switch off makes a gap in the circuit, so the electricity stops flowing.

Mains electricity in your home flows through live or 'phase' wires linked to every light, socket outlet and fixed electrical appliance in your home. The current flows back out of the building through the neutral wires.

Earthing

Any material through which electricity can flow is known as a conductor. Most metals conduct electricity well, which is why copper is used for electrical wiring. The earth itself – the ground on which we stand – is also an extremely good conductor, which is why electricity always flows into the earth whenever it can, taking the shortest available route. This means that if you were to touch a live conductor, the current would divert and take the short route to the earth – through your body.

A similar thing can happen if a live wire accidentally comes into contact with any exposed metal component of an appliance, including its casing. To prevent this, a third wire is included in the system and connected to the earth, usually via the outer casing of the electricity company's main service cable. This third wire – called the earth wire – is attached to the metal casing of some appliances and to earth terminals in others, providing a direct route to the ground should a fault occur. This sudden change of route by the electricity – known as an earth fault – causes a fuse to blow or circuit breaker to operate, cutting off the current.

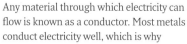

Double insulation
Appliances that are double-insulated – which usually means they have a plastic casing that insulates the user from metal parts that could become live – must not be earthed with a third wire. A square within a square, either printed or moulded on an appliance, means it is double-insulated and its flex does not need an earth wire.

Notifiable work
To help you decide which electrical work you can do yourself, all notifiable work shown in this book is marked with this symbol.

Measuring electricity

Watts measure the amount of power used by an appliance when working. The wattage of an electrical appliance is normally marked on its casing. One thousand watts (1000W) equal one kilowatt (1kW).

Amps measure the flow of current that is necessary to produce the required wattage for an appliance.

Volts measure the 'pressure' provided by the electricity company. This drives the current along the conductors to the various outlets. In this country 230 volts (formerly 240) is standard.

If you know two of these measurements, you can determine the other one.

$\dfrac{\text{Watts}}{\text{Volts}}$ = Amps	Amps x Volts = Watts
Use this method to determine what kind of fuse or flex is safe.	Indicates how much power is needed to operate an appliance.

Building Regulations on electrical wiring

Regulations – known as Part P of the Building Regulations – have been introduced to promote better standards. Similar legislation for Scotland is covered by the Building Standards (Scotland) Regulations. These regulations do not prevent the DIY worker from undertaking electrical wiring, but they put strict limits on what can be done without supervision and inspection.

All local authorities have Building Control Officers (BCOs) who are responsible for monitoring the regulations. You should therefore contact your BCO to ascertain how exactly your particular authority applies the regulations.

Certain tasks can be undertaken without having to notify the authority, and most BCOs do not require any communication or paperwork for 'non-notifiable' electrical work. However, where similar work is carried out in locations such as kitchens and bathrooms or the wiring will be complicated or extensive, then the work becomes 'notifiable' and must be discussed with the BCO before it is carried out.

Non-notifiable work

A DIY worker can do the following, anywhere in the home without having to inform the BCO in advance:
- Replace sockets, fused connection units, switches and ceiling roses.
- Replace damaged cable for a single circuit.
- Refix or replace enclosures (mounting boxes for sockets and switches) on existing circuits.
- Provide mechanical protection in the form of conduit and plastic channel.

Also without having to inform the BCO in advance, a DIY worker can do the following anywhere in the home except in kitchens, bathrooms and utility rooms and in special locations – such as rooms with a bathtub or shower, swimming pools and paddling pools, hot-air saunas and outdoors – or when installing extra-low-voltage lighting:
- Add new light fittings and switches to existing circuits.
- Add new socket outlets to existing ring circuits and radial circuits.
- Add fused spurs to existing ring circuits and radial circuits.

Notifiable work

The BCO must be notified before a DIY worker undertakes any electrical work not listed above or if the work is categorized as one of the exceptions described above.

Although DIY electricians are permitted to carry out notifiable work, the cost and complexity of obtaining approval, testing, and certification from the BCO (see opposite) may make it more economical to have such work done by a professional electrician.

SEE ALSO > Testing circuits 74–5, Switching off the power 76, Earthing and bonding 77

Complying with the regulations

It is not necessary to involve the BCO when any work is undertaken by a professional electrician – that is a competent person registered with an electrical self-certification scheme. He or she will deal with all the paperwork required by the authority and, on completion of the work, should give you a signed Building Regulations Self-certification Certificate and a completed Electrical Installation Certificate.

If you feel competent to do any notifiable work yourself, you must tell your local BCO in advance exactly what you propose to do. Having obtained permission to proceed, once you have completed the work you must ask the BCO to send an inspector to test the installation and issue a certificate. A fee will be charged for inspection and testing. Most BCOs will offer some concession – such as including the electrical inspection in the general inspection costs for building a new extension. Some BCOs may ask you to arrange for a competent electrician to inspect and test the work.

Procedures are laid down for appeals, determinations, relaxations and dispensations, but the common-sense approach is to accept any advice or instruction given by your local Building Control Officer.

Provided you are competent to undertake non-notifiable work, you may proceed without supervision so long as the methods used comply with the the IEE Wiring Regulations (BS7671). Good workmanship and the use of proper materials are fundamental to these regulations. You should also keep a record of your work in the form of a Minor Electrical Works Certificate, which you can pass on to interested parties should you decide to sell your home in the future. Selling a house without appropriate paperwork may introduce delays and difficulties.

The methods and materials suggested in this book comply with the Wiring Regulations, but you should be aware that the rigorous final testing of installations, which has to be carried out by qualified persons using specialized equipment, falls outside the scope of this book. If you have any doubts about your ability to satisfy the requirements of the Wiring Regulations, then use this book to study the work involved so you can brief a professional electrician and agree an appropriate price for the job. Ensure that any electrician you hire is a member of an authorized competent-person self-certification scheme.

With safety in mind

Throughout this chapter you will find frequent references to the need for safety while working on your electrical system – but it cannot be stressed too strongly that you must also take steps to safeguard yourself and anyone else using the system. Faulty wiring and appliances are dangerous, and can be lethal.

- Never inspect or work on any part of an electrical installation without first switching off the power at the consumer unit and removing the relevant circuit fuse or locking off the miniature circuit breaker (MCB).
- Always unplug a portable appliance or light before doing any work on it.
- Always use the correct tools and use good-quality equipment and materials.
- Always double-check all your work (especially connections) before you turn the electricity on again.
- Fuses are vital safety devices. Never fit one that's rated too highly for the circuit it is to protect – and never be tempted to use any other type of wire or metal strip in place of proper fuses or fuse wire.
- Wear rubber-soled shoes when you're working on an electrical installation.

Colour coding

For purpose of identification, the coverings of live, neutral and earth wires in flex and mains cables are colour-coded. The live wires are brown and the neutral wires blue.

When an earth wire is included in a piece of flex, it is coded with a green-and-yellow covering. In a mains cable, the earth is a bare copper wire. Whenever the earth wire is exposed for linking to socket outlets or light fittings, it should be

covered with a green-and-yellow sleeve.

Until recently the live and neutral wires in mains cables were coded with other colours. Live wires had a red covering, and neutral wires a black covering. There is no reason to replace old colour-coded cables, but whenever you need to join new cable to old, remember to connect the brown wire to the old red one, and the blue wire to the old black one.

Identifying conductors
The insulation used to cover the conductors in electrical cable and flex is colour-coded to indicate live, neutral and earth.

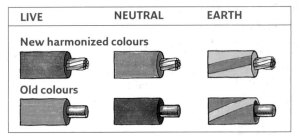

LIVE	NEUTRAL	EARTH
New harmonized colours		
Old colours		

Fuses and circuit breakers

A conductor will heat up if an unusually powerful current flows through it. This can damage electrical equipment and create a serious fire risk if it is allowed to continue. As a safeguard weak links are included in the wiring, to break the circuit before the current reaches a dangerously high level.

Fuses

A very common form of protection is a fuse, a thin wire that's designed to break the circuit by melting at a specific current. This varies according to the part of the system that the fuse is protecting – an individual appliance, a single power or lighting circuit, or the entire domestic wiring system.

Miniature circuit breakers

Alternatively, a special switch called a circuit breaker is used that trips and cuts off the current as soon as an overload on the wiring is detected.

A fuse will 'blow' (or an MCB will trip) in the following circumstances:
- If too many appliances are operated on a circuit simultaneously, then the excessive demand for electricity will blow the fuse in that circuit.
- If the current reroutes to earth due to a faulty appliance, the flow of power increases in the circuit and blows the fuse (this is known as an earth fault).

SEE ALSO > Switching off the power 76

Bathroom safety

Because water is such a highly efficient conductor of electric current, water and electricity form a very dangerous combination. For this reason, in terms of electricity, bathrooms are potentially the most dangerous areas in your home. Where there are so many exposed metal pipes and fittings, combined with wet conditions, regulations must be stringently observed if fatal accidents are to be avoided.

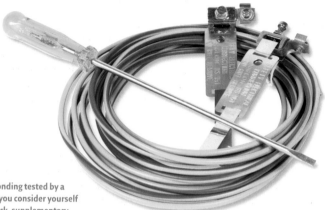

N If you undertake work marked with this symbol, you must inform the BCO before starting – see FIRST THINGS FIRST.

WARNING
Have your supplementary bonding tested by a qualified electrician. Unless you consider yourself fully competent to do the work, supplementary bonding must be installed by a professional.

GENERAL SAFETY

- Sockets must not be fitted in a bathroom - except for special shaver sockets that conform to BS EN 60742 Chapter 2, Section 1.

- The IEE Wiring Regulations stipulate that light switches in bathrooms must be outside zones 0 to 3 (see opposite). The best way to comply with this requirement is to fit ceiling-mounted pull-cord switches.

- When installing an electrical appliance in a bathroom, the relevant circuit should be protected by a 30 milliamp RCD.

- If you have a shower in a bedroom, it must be not less than 3m (9ft 11in) from any socket outlet, which must be protected by a 30 milliamp RCD.

- Light fittings in a bathroom must be well out of reach and shielded - so fit a close-mounted ceiling light, properly enclosed, rather than a pendant fitting.

- Never use a portable appliance, such as a hairdryer, in a bathroom - even if it is plugged into a socket outside the room.

Supplementary bonding **N**

In any bathroom there are many nonelectrical metallic components, such as metal baths and basins, supply pipes to bath and basin taps, metal waste pipes, radiators, central-heating pipework and so on - all of which could cause an accident during the time it would take for an electrical fault to blow a fuse or trip a miniature circuit breaker. To ensure that no dangerous voltages are created between metal parts, all these metal components must be connected one to another by a conductor which is itself connected to a terminal on the earthing block in the consumer unit. This is known as supplementary bonding and is required for all bathrooms - even when there is no electrical equipment installed in the room, and even though the water and gas pipes are bonded to the consumer's earth terminal near the consumer unit.

When electrical equipment, such as a heater or shower, is fitted in a bathroom, that too must be bonded by connecting its metalwork - such as the casing - to the nonelectrical metal pipework.

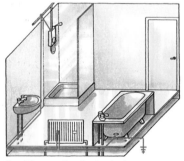
Supplementary bonding in a bathroom

Making the connections

The Wiring Regulations specify the minimum size of earthing conductor that can be used for supplementary bonding in different situations, so that large-scale electrical installations can be costed economically. However, 6mm² single-core cable, insulated with green-and-yellow PVC, is large enough to be safe for supplementary bonding in any domestic situation. For a neat appearance, the route of the bonding cable should be planned to run behind the bath panel, under floorboards, and through basin pedestals. If necessary, the cable can be run through a hollow wall or under plaster.

Connecting to pipework
An earth clamp is used for making connections to pipework. The pipe should be cleaned locally with wire wool to make a good connection between the pipe and clamp. Scrape or strip an area of paintwork if the pipe has been painted.

Attach an earth clamp to the pipework

Connecting to a bath or basin
Metal baths or basins are made with an earth tag. To connect the earth cable, the bared end of the conductor should be trapped under a nut and bolt with metal washers. Make sure the tag has not been painted or enamelled.

If an old metal bath or basin has not been provided with an earth tag, drill a hole through the foot of the bath or through the rim at the back of the basin; so that the cable can be connected with a similar nut and bolt, with metal washers.

Connecting to an appliance
The bonding cable must be connected to the earth terminal provided in the electrical appliance and then run to a clamp on a metal supply pipe nearby.

Zones for bathrooms

Within a room containing a bath or shower, the Wiring Regulations define areas, or zones, where specific safety precautions apply. The regulations also describe what type of appliances can be installed in each zone, and the routes cables must take. There are special considerations for extra-low-voltage equipment with separated earth.

The four zones

Any room containing a bathtub or shower is divided into four zones. Zone 0 is the interior of the bathtub or shower tray – not including the space beneath the tub, which is covered by other regulations (see right). Zones 1 to 3 are specific areas above and around the bath or shower, where only specified electrical appliances and their cables may be installed. Wiring outside these areas must conform to the IEE Wiring Regulations, but no specific 'zone' regulations apply.

ZONE	LOCATION	PERMITTED
Zone 0	Interior of the bathtub or shower tray.	No electrical installation.
Zone 1	Directly above the bathtub or shower tray, up to a height of 2.25m (7ft 5in) from the floor. (See also UNDER THE BATH, right.)	Instantaneous water heater. Instantaneous shower. All-in-one power shower, with a suitably waterproofed integral pump. The wiring that serves appliances within the zone.
Zone 2	Area within 0.6m (2ft) horizontally from the bathtub or shower tray in any direction, up to a height of 2.25m (7ft 5in) from the floor. The area above zone 1, up to a height of 3m (9ft 11in) from the floor.	Appliances permitted in zone 1. Light fittings. Extractor fan. Space heater. Whirlpool unit for the bathtub. Shaver socket to BS EN 60742 Chapter 2, Section 1. The wiring that serves appliances within the zone and any appliances in zone 1.
Zone 3	Up to 2.4m (7ft 11in) outside zone 2, up to a height of 2.25m (7ft 5in) from the floor. The area above zone 2 next to the bathtub or shower, up to a height of 3m (9ft 11in) from the floor.	Appliances permitted in zones 1 and 2. Any fixed electrical appliance (a heated towel rail, for example) that is protected by a 30 milliamp RCD. The wiring that serves appliances within the zone and any appliances in zones 1 and 2.

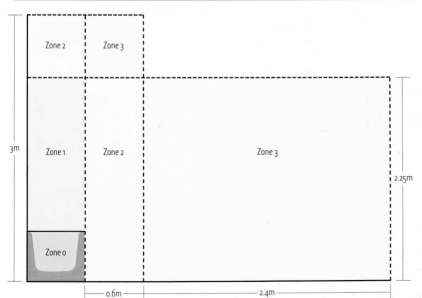

Zones within a room containing a bath or shower

Under the bath

The space under a bathtub is designated as zone 1 if it is accessible without having to use a tool – that is, if there is no bath panel or if the panel is attached with magnetic catches or similar devices that allow the panel to be detached without using a tool of some kind. If, however, the panel is screw-fixed, then the enclosed space beneath the bath is considered to be outside all zones.

Supplementary bonding

In bathrooms, nonelectrical metallic components must be bonded to earth (see opposite). In zones 1, 2 and 3, this bonding is required to all pipes, any electrical appliances and any exposed metallic structural components of the building. This does not include window frames, unless they are themselves connected to metallic structural components.

Supplementary bonding is not required outside the zones. In the case of a shower cubicle in a bedroom, supplementary bonding can also be omitted from zone 3.

Switches

Electrical switches, including ceiling-mounted switches operated by a pull cord, must be situated outside the zones. The only exceptions are those switches and controls incorporated in appliances suitable for use in the zones.

If the bathroom ceiling is higher than 3m (9ft 11in), pull-cord switches can be mounted anywhere. However, if the ceiling height is between 2.25 and 3m (7ft 5in and 9ft 11in), pull-cord switches must be mounted at least 0.6m (2ft) – measured horizontally – from the bathtub or shower cubicle. If the ceiling is lower than 2.25m (7ft 5in), switches must be outside the room.

IP coding

Electrical appliances installed in zones 1 and 2 must be made with suitable protection against splashed water. This is designated by the code IPX4 (the letter X is sometimes replaced with a single digit). Any number larger than four is also acceptable as this indicates a higher degree of waterproofing. If in doubt, check with your supplier.

● **Cable runs**
You are not permitted to run electrical cables that are feeding a zone through another zone designated with a lower number. This includes cables buried in the plaster or concealed behind other wallcoverings.

● **13amp sockets**
In the special case of a bedroom containing a shower cubicle, socket outlets are permitted in the room, but only outside the zones, and the circuit that feeds the sockets must be protected by a 30 milliamp RCD.

IP coding
Suitable equipment may be marked with this symbol.

SEE ALSO > Wiring a shower 81

Testing circuits

For your own safety and for the safety of others using or working on your electrical installation in the future, carry out the tests recommended here each time you work on the fixed wiring of your home. Although these tests should not be regarded as substitutes for those required by Building Control Officers, if your work passes them it is likely to meet the requirements of the regulations.

Test instruments

- **Ohms and MegOhms**
On some multimeters, Ohms are indicated by the symbol Ω and MegOhms (millions of Ohms) by MΩ.

There are many different testers to choose from, and those at the top end of the market may cost hundreds of pounds. However, reliable testers can be purchased for much less.

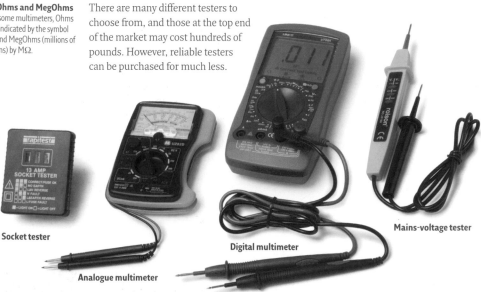

Socket tester

Analogue multimeter

Digital multimeter

Mains-voltage tester

Mains-voltage tester
Mains testers are designed to tell you whether a circuit is completely 'dead' after you have turned off the power at the consumer unit. Be sure to buy a tester that's intended for use with mains voltage – similar devices are sold in auto shops for 12V car wiring only.

Multimeters
There are digital multimeters that give readings on an LCD screen, and there are analogue instruments with a needle that moves across a scale on a dial.

Both types have a rotary selector for setting the meter to measure voltage, current, resistance and other values. When testing insulation resistance and continuity (see opposite), you want the meter to measure resistance. For insulation resistance, set the selector to the highest resistance range (**1**); for continuity, set it to the lowest resistance range (**2**). With some multimeters, you can select an audible signal to tell you when there is continuity.

With an analogue multimeter, a reading at the lower end of the scale indicates a

continuous, unbroken circuit. A reading at the top end of the scale means there is an 'open' circuit – there's a break somewhere on the circuit.

With an analogue meter, satisfactory insulation resistance is indicated when the needle is at the top end of the scale, showing a reading of millions of Ohms (MegOhms).

When using a digital multimeter, for continuity you are looking for a low value – less than 10 Ohms. For satisfactory insulation resistance, you want a value of more than 10 million Ohms (10 MegOhms).

To test that a multimeter is working, touch the two probes together and the meter should read zero. Move the probes apart and it should read infinity. The manufacturers of digital multimeters use the figure 1 (with no other figures after the decimal point) to mean infinity – which simply indicates a very high resistance.

Socket tester
A simple plug-in device can be used to test the connections inside a 13amp socket outlet without having to turn off the power or expose internal wiring.

1 Insulation resistance
Set the selector to the highest resistance scale.

2 Continuity
Set the selector to the lowest resistance scale.

Required testing

Carry out the tests described opposite to check the effectiveness of the fixed wiring of your house. However, the Building Regulations require final testing with specialist equipment, which is used by Building Control inspectors and other professionals to provide readings that can be entered on the necessary certificates.

Is the power off?

Having turned off the power at the consumer unit, make sure an accessory is safe to work on by using an electronic mains-voltage tester to check whether terminals or wires are live, before you tamper with them. Always make sure the tester is functioning properly both before and after you use it, by testing it on a circuit you know to be live.

Following the manufacturer's instructions, place one probe on the neutral terminal and the other on the live terminal; if the indicator lights up, the circuit is live. If it doesn't illuminate, test again – this time between the earth terminal and each of the live and neutral terminals. If the indicator still doesn't light up, you can assume the circuit is not live – provided you have checked the tester.

Using a mains-voltage tester
Touch the neutral terminal with one probe and the live terminal with the other. The circuit is live if the indicator illuminates.

Protective devices

As part of the testing required by Building Control Officers, an inspector uses equipment that checks whether protective devices (miniature circuit breakers and residual current devices) are operating within the times specified in the regulations. This is a procedure that falls outside the scope of this book.

However, you can at least ensure an RCD trips satisfactorily when you press its test button, and you can test that an MCB is working by operating its switch. Test these devices every time you carry out electrical work on your home, and also at regular intervals – say every three months or so.

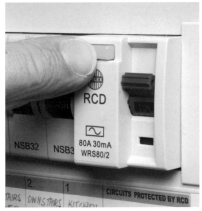

Press the test button on a residual current device

Checking polarity

Having worked on an electrical circuit, before replacing sockets, switches or faceplates, always double-check that the live, neutral and earth conductors are connected to their respective terminals.

Having switched the power back on, you can use a plug-in tester to check the wiring to your socket outlets. Switch on the socket; and if all three indicator lights come on, the socket is wired correctly.

If all three lights come on, polarity is correct

Testing for continuity

In simple terms, a continuity test is designed to check that there is an electrical connection between two points, say between the two ends of a length of wire. This is done with a multimeter set to its lowest resistance range. Place one of the instrument's probes on each end of the wire. A low reading on the dial or screen means the wire is continuous – there is continuity. A high reading means the wire is broken somewhere along its length – so there's no continuity.

You can use this test to check whether a cartridge fuse has blown, or whether a ring circuit is continuous or a heating element is working.

Testing a ring circuit

This is an important test because a ring circuit could still be functioning even when there's a break somewhere on the circuit. At the consumer unit, having first turned off the main switch, identify the two ends of the cable feeding the ring circuit; then remove the live, neutral and earth conductors from their terminals and separate all the wires.

Testing insulation resistance

This test is designed to make sure there is no current leaking through the insulation between two conductors, say live and neutral. If this is allowed to happen there is a danger of overheating – which could cause a fire. Alternatively, there could be a short circuit and the fuse protecting the circuit would blow or the MCB would trip.

You should test any circuit you have worked on. Make sure the power is turned off, then unplug every appliance on the circuit and check that all switches are turned off. This includes switches for all fixed appliances.

At the consumer unit, identify the cable feeding the relevant circuit and disconnect the conductors from their terminals. If you are installing a new circuit, do the test before making the final connections at the consumer unit.

With a multimeter set to its highest resistance range, place one probe on the live (red or brown) conductor and the other probe on the neutral (black or blue) conductor. If the reading shown on the meter is low, the insulation resistance is suspect and should be investigated. If the reading is high – in the order of MegOhms – the insulation resistance is satisfactory. Repeat the test between the live conductor and earth (green-and-yellow) conductor

Testing for continuity
Touch one end of the wire with one probe and the other end of the same wire with the other probe. A low reading (shown above) indicates that the wire is continuous.

Carry out the continuity test between the two ends of the live (red or brown) conductor. Repeat the test on the neutral (black or blue) conductor, and then on the earth (yellow-and-green) conductor. A low reading (low resistance) for each test means there is continuity on all three conductors – each is continuous, which is how it should be.

If you get a high reading, check every socket outlet, junction box and fused connection unit to make sure there are no loose connections and then perform the continuity test again.

Insulation resistance
Touch the neutral conductor with one probe and the live conductor with the other. A high reading (shown above) indicates satisfactory insulation resistance.

and then between the neutral and earth, looking for a high reading in each case.

Investigating a suspect circuit

With the power turned off, inspect the suspect wiring, including every accessory (socket, switch, junction box and so on), starting with the ones you have been working on. Typical causes of low insulation resistance are:
- A damp wall or water running into an accessory.
- A nail or screw driven through a cable.
- Rodent damage to cables.
- Old rubber-insulated cables connected to the circuit.
- Conductors crushed together by careless replacement of a socket outlet or switch.

SEE ALSO > Building Regulations on electrical wiring 70, RCDs 77, Replacing an element 82

Main switch equipment

Electricity flows because of a difference in 'pressure' between the live wire and the neutral one, and this difference in pressure is measured in volts. Domestic electricity in this country is supplied as alternating current, at 230 volts, by way of the electricity company's main service cable. This normally enters your house underground, although in some areas electricity is distributed by overhead cables.

The service head

The main cable terminates at the service head, or 'cutout', which contains the service fuse. This fuse prevents the neighbourhood's supply being affected if there should be a serious fault in the circuitry of your house. Cables connect the cutout to the meter, which registers how much electricity you consume. Both the meter and cutout belong to the electricity company and must not be tampered with. The meter is sealed in order to disclose interference.

If you are using cheap night-time power for electric storage heaters and hot water, a time switch will be supplied by the electricity company.

Consumer unit

Electricity is fed to and from the consumer unit by 'meter leads', thick single-core insulated-and-sheathed cables made up of several wires twisted together. The consumer unit is a box containing the fuseways that protect the individual circuits in the house. It also incorporates the main isolating switch, which you operate when you need to cut off the supply of power to the whole house. Not all main isolating switches operate the same way. Before you need to use it, check to see whether the main switch on your consumer unit has to be in the up or down position for 'off'.

In a house where several new circuits have been installed over the years, the number of circuits may exceed the number of fuseways in the consumer unit. If so, an individual switchfuse unit – or more than one – may have been mounted alongside the main unit. Switchfuse units comprise a single fuseway and an isolating switch; they, too, are connected to the meter by means of meter leads.

If your home is heated by off-peak storage heaters, then you will have an Economy 7 meter and a separate consumer unit for the heater circuits.

● Cross-bonding cable sizes
Single-core cables are used to cross-bond gas and water pipes to earth. An electrician can calculate the minimum size for these cables, but for any single house or flat it is safe to use 10mm² cable. (See also PME opposite).

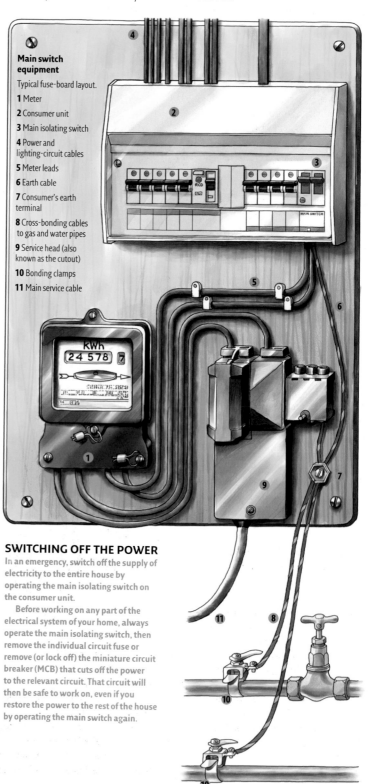

Main switch equipment

Typical fuse-board layout.

1 Meter
2 Consumer unit
3 Main isolating switch
4 Power and lighting-circuit cables
5 Meter leads
6 Earth cable
7 Consumer's earth terminal
8 Cross-bonding cables to gas and water pipes
9 Service head (also known as the cutout)
10 Bonding clamps
11 Main service cable

SWITCHING OFF THE POWER

In an emergency, switch off the supply of electricity to the entire house by operating the main isolating switch on the consumer unit.

Before working on any part of the electrical system of your home, always operate the main isolating switch, then remove the individual circuit fuse or remove (or lock off) the miniature circuit breaker (MCB) that cuts off the power to the relevant circuit. That circuit will then be safe to work on, even if you restore the power to the rest of the house by operating the main switch again.

SEE ALSO ▶ Fuses and circuit breakers 71, Supplementary bonding 72

Earthing and bonding

All of the individual earth conductors of the various circuits in the house are connected to a metal earthing block in the consumer unit. A single cable with a green-and-yellow covering runs from this earthing block to the consumer's earth terminal, which is mounted next to the cutout. In most urban houses a connection is provided from inside the cutout to an external earth-connection block, which is also wired to the consumer's earth terminal. This provides an effective path to earth, as it allows the current to pass along the sheath of the main service cable to the electricity company's substation, where it is solidly connected to earth. If the company does not provide an earth connection to its cable, the installation must be protected by an RCD.

In the past most domestic electrical systems were earthed to the cold-water supply, so earth-leakage current passed out along the metal water pipes into the ground in which they were buried. But nowadays more and more water systems use nonconductive, nonmetallic pipes and fittings. As a result, such a means of earthing is no longer reliable.

Despite this, you will find that your gas and water pipework is connected to the consumer's earth terminal. This ensures that the water and gas piping systems are cross-bonded, so earth-leakage current passing through either system will run without hindrance to the main earth without producing dangerously high voltages. The cross-bonding clamps must be as close as possible to the point where the pipes enter the house, but on the consumer's side (within 600mm) of the stopcock or gas meter.

Bonding clamp
This type of clamp (BS 951) is used to make connections to gas and water pipes. It must not be removed under any circumstances.

PME

The electricity company sometimes provides a different method of earthing the system, called 'protective multiple earth' (PME), by which earth-leakage current is fed back to the substation along the neutral return wire, and so to earth. Regulations regarding the earthing of this system are particularly stringent. With PME, cross-bonding cables to gas and water services are sometimes required to be larger. Check this with the electricity company.

RCDs

Although the local electricity company normally provides effective earthing for the electrical system of your home, safe earthing is the consumer's own responsibility. With this in mind, it is worth installing a residual current device (RCD) into the house circuitry.

When conditions are normal, the current flowing out through the neutral conductor is exactly the same as that flowing in through the live one. Should there be an imbalance between the two caused by an earth leakage, the residual current device will detect it immediately and isolate the circuitry.

An RCD can be either installed as a

A separate unit containing an RCD

separate unit or incorporated into the consumer unit, sometimes together with the main isolating switch.

A residual current device is sometimes referred to as a residual current circuit breaker (RCCB). It was formerly known as an ELCB, or earth-leakage circuit breaker.

Old fuse boards

Domestic wiring systems were once very different from the ones used today. Besides lighting, water-heating and cooker circuits, each socket outlet had its own circuit and fuse, while further circuits would be installed from time to time as the needs of the household changed. Consequently, an old house may have a mixture of 'fuse boxes' attached to the fuse board, along with the meter.

You may find that the wiring itself is haphazard and badly labelled, with the serious danger that you may not safely isolate a circuit you're going to work on. Furthermore, you will not be able to tell whether a particular fuse is correctly and safely rated unless you know what type of circuit it is protecting.

Arrange for an inspection

If your home still has such an old-style fuse board, have it inspected and tested by a qualified electrician before you attempt to work on any part of the system. He or she can advise you as to whether your installation needs to be replaced with a modern consumer unit. At the same time, check that the cables are PVC-insulated. If everything does prove to be in good working condition, he or she can label the various circuits clearly to help you in the future.

An old-fashioned fuse board
This type of installation is out of date. A professional electrician may advise you to replace at least some of the components.

SEE ALSO > Bathroom safety 72

Fixed appliances

Socket outlets are designed to enable portable appliances to be moved from room to room, but many electrical appliances are fixed to the structure of the house or stand in one position all the time. Such appliances may therefore just as well be wired into your electrical installation permanently. Indeed in some cases there is no alternative, and some require radial circuits of their own direct from the consumer unit.

Fused connection units

A fused connection unit (FCU) is basically a device for joining the flex (or sometimes cable) of an appliance to circuit wiring. The connection unit incorporates the added protection of a cartridge fuse similar to that found in a 13amp plug. If the appliance is connected by a flex, choose a unit that has a cord outlet in the faceplate.

Some fused connection units are fitted with a switch, and some of these have a neon indicator that shows at a glance whether they are switched on. A switched connection unit allows you to isolate the appliance from the mains.

All fused connection units are single (there are no double versions available) and have square faceplates that fit metal boxes for flush mounting or standard surface-mounted plastic boxes.

Changing a fuse
With the electricity turned off, remove the retaining screw in the face of the fuse holder. Take the holder from the connection unit; prise out the old fuse and fit a new one; then replace the holder and the retaining screw.

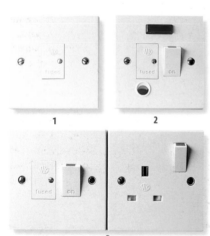

Fused connection units
1 Unswitched connection unit.
2 Switched unit with cord outlet and indicator.
3 Connection unit and socket outlet in a dual mounting box.

Mounting a fused connection unit

A fused connection unit is mounted in the same type of box as an ordinary socket outlet, and the box is fixed to the wall in exactly the same way. The unit can also be mounted in a dual box that is designed to hold two single units – for example, a standard socket outlet beside a connection unit. The socket is wired to the ring circuit, and the two units are linked together inside the box by a short 2.5mm² spur.

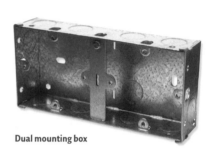

Dual mounting box

Wiring a fused connection unit

Before you wire a fused connection unit to the house circuitry, the power must be switched off at the consumer unit.

Fused connection units can be supplied by a ring circuit, a radial circuit or a spur. Some appliances are connected to the unit with flex, others with cable. Either way, the wiring arrangement inside the unit is the same. Units with cord outlets have clamps to secure the connecting flex.

An unswitched connection unit has two live (L) terminals – one marked 'Load' for the brown wire of the flex, and the other marked 'Mains' for the brown or red wire from the circuit cable. The blue wire from the flex and the blue or black wire from the circuit cable go to similar neutral (N) terminals; and both earth wires are connected to the unit's earth (E) terminal or terminals (**1**).

Switched connection units

A fused connection unit with a switch has two sets of terminals, too. Those marked 'Mains' are for the spur or ring cable that supplies the power; the terminals marked 'Load' are for the flex or cable from the appliance.

Wire up the flex side first, connecting the brown wire to the L terminal, and the blue one to the N terminal, both on the 'Load' side. Connect the green-and-yellow wire to the E terminal (**2**) and tighten the cord clamp.

Attach the circuit conductors to the 'Mains' terminals – brown or red to L, and blue or black to N; then sleeve the earth wire and take it to the E terminal (**2**).

If the fused connection unit is on a ring circuit, you need to fit two circuit conductors into each 'Mains' terminal and into the earth terminal.

Before screwing the unit to its box, make sure the wires are held firmly in the terminals and can fold away neatly.

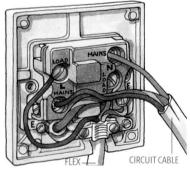

1 Wiring a fused connection unit

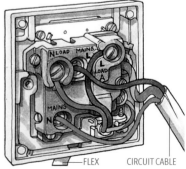

2 Wiring a switched fused connection unit

SEE ALSO > Switching off the power 76

Wiring small fixed appliances

Small permanent electrical appliances with ratings of up to 3000W (3kW) – wall heaters, extractor fans, cooker hoods and so on – can be wired into a ring or radial circuit by means of fused connection units. Although such appliances could be connected by means of 13amp plugs to socket outlets, the electrical contact would not be so good – and there is also some risk of fire with that type of permanent installation.

Flex outlets

Flexible-cord outlet

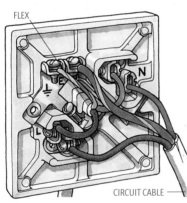 If you undertake work marked with this symbol, you must inform the BCO before starting – see FIRST THINGS FIRST.

In some situations, such as a bathroom, the fused connection unit has to be mounted in a different location from the appliance it is supplying with power.

If the appliance is fitted with flex, you can mount a flexible-cord outlet – 'flex outlet' next to the appliance – and then run a cable from the outlet to the FCU and connect it to the 'Load' terminals in the unit.

The flex outlet is mounted either on a standard surface-mounted box or flush on a metal box. At the back of the faceplate are three pairs of terminals to take the conductors from the flex and the cable.

FLEX

CIRCUIT CABLE

Wiring a flexible-cord outlet

Extractor fans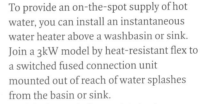

To install a fan in a kitchen, mount a fused connection unit 150mm (6in) above the worktop. Run a cable to the fan or to a flexible-cord outlet next to it. If the fan has no integral switch, use a switched connection unit to control it. Fit a 3 or 5amp fuse, as recommended by the manufacturer.

If the fan's speed and direction are controllable, it may have a separate control unit – in which case you need to wire the connection unit to the control unit, following the manufacturer's instructions.

To install a fan in a bathroom, mount the fused connection unit outside the room and run the cable to the fan or flex outlet via a double-pole ceiling switch. The fan must be outside zones 0 and 1, and the circuit protected by a 30 milliamp RCD.

Wall-mounted fan
Run a 1.5mm² cable from a fused connection unit to a wall-mounted extractor fan.

Fridges, dishwashers and washing machines

There is no reason why you cannot plug an appliance like a fridge, dishwasher or washing machine into a standard socket outlet – except that in a modern kitchen such appliances are installed under worktops, and sockets mounted behind them are difficult to reach. It's therefore generally more convenient to mount a switched fused connection unit 150mm (6in) above the worktop, then connect it to the ring circuit and run a spur – using 2.5mm² cable – from the connection unit to a socket outlet mounted behind the appliance.

Cooker hoods

Either mount a fused connection unit, fitted with a 3amp fuse, close to the cooker hood or mount the connection unit 150mm (6in) above the worktop and then run a 1.5mm² cable from the unit to a flexible-cord outlet beside the hood.

Instantaneous water heaters

To provide an on-the-spot supply of hot water, you can install an instantaneous water heater above a washbasin or sink. Join a 3kW model by heat-resistant flex to a switched fused connection unit mounted out of reach of water splashes from the basin or sink.

If the heater is for use in a bathroom, wire it via a flex outlet to a ceiling pull-switch and then to a fused connection unit outside the bathroom. Fit a 13amp fuse in the unit.

Wire a 7kW water heater in the same way as a shower. If it is situated in the kitchen, you can use a double-pole wall switch to control it.

Waste-disposal units

A waste-disposal unit is housed in the cupboard unit below the sink. Mount a switched fused connection unit 150mm (6in) above a worktop near the sink, but well out of reach of small children and splashes from the sink. From the unit, run a 1.5mm² cable to a flex outlet next to the waste-disposal unit. Clearly label the connection unit 'waste disposal', to avoid accidents. Fit a 13amp fuse.

Circuits for kitchen equipment
1 Connection units
2 Flex outlets
3 Socket outlets

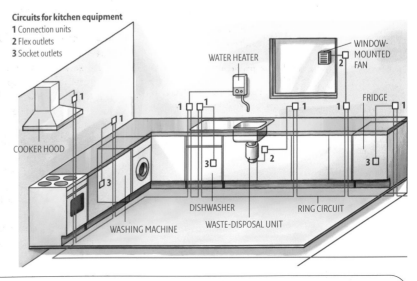

WATER HEATER

WINDOW-MOUNTED FAN

FRIDGE

COOKER HOOD

WASHING MACHINE

DISHWASHER

WASTE-DISPOSAL UNIT

RING CIRCUIT

SEE ALSO ▶ Building Regulations on electrical wiring 70, Bathroom safety 72, Zones for bathrooms 73, Switching off the power 76, Double-pole switches 81, Waste-disposal units 48

N If you undertake work marked with this symbol, you must inform the BCO before starting – see FIRST THINGS FIRST.

Shaver sockets **N**

Special shaver socket outlets are the only kind of electrical socket allowed in bathrooms. This type of socket contains a transformer that isolates the user side of the unit from the mains, reducing the risk of an electric shock.

This type of socket has to conform to the exacting British Standard BS EN 60742 Chapter 2, Section 1. However, there are shaver sockets that do not have an isolating transformer and therefore don't conform to this standard. These are safe to install and use in a bedroom – but must not be fitted in a bathroom.

• **RCD protection**
When installing any electrical appliance in a bathroom, the circuit must be protected by a 30 milliamp RCD.

Shaver unit for use in a bathroom

You can wire a shaver socket from a junction box on an earthed lighting circuit or from a fused connection unit, fitted with a 3amp fuse, on a ring-circuit spur. If you're installing the shaver socket in a bathroom, then the fused connection unit must be positioned outside the room. Run a 1.5mm² two-core-and-earth cable from the connection unit to the shaver socket; then connect the conductors: brown to L and blue to N. Sheathe the earth wire with a green-and-yellow sleeve and connect it to the earth (E) terminal.

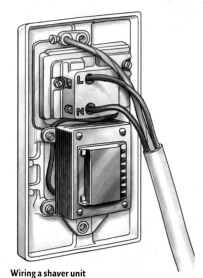

Wiring a shaver unit

Heated towel rails **N**

When you're installing a heated towel rail in a bedroom, the appliance can be wired to a switched fused connection unit mounted next to the appliance, at a height of about 225 to 300mm (9in to 1ft) from the floor. For a towel rail of 1kW or less, fit a 5amp fuse in the connection unit; otherwise fit a 13amp fuse.

Heated towel rail in a bathroom
The Wiring Regulations covering other electrical equipment situated in bathrooms also apply to heated towel rails. The fused connection unit must be mounted outside the bathroom and wired to a convenient mains power circuit. Because the towel rail is mounted inside the bathroom, the circuit supplying power must be protected by a 30 milliamp RCD.

Run a spur cable from the fused connection unit to a flexible-cord outlet mounted beside the towel rail. Connect the cable's brown conductor to the flex outlet's live (L) terminal, and the blue conductor to the neutral (N) terminal.

Cover the bare copper earth wire with green-and-yellow sleeving and then attach it to the outlet's earth terminal.

Connect the flex from the towel rail similarly – brown conductor to live and blue to neutral. Connect the green-and-yellow-insulated earth conductor to the earth terminal.

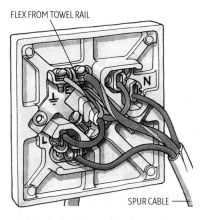

FLEX FROM TOWEL RAIL

SPUR CABLE

Wiring the flex outlet in a bathroom

Electric towel rail installed in a stylish modern bathroom

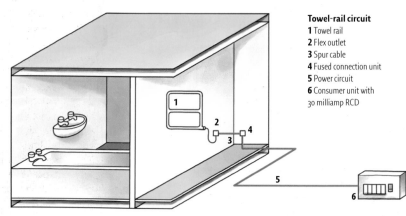

Towel-rail circuit
1 Towel rail
2 Flex outlet
3 Spur cable
4 Fused connection unit
5 Power circuit
6 Consumer unit with 30 milliamp RCD

SEE ALSO > Building Regulations 70, Bathroom safety 72, Zones for bathrooms 73, Switching off 76, Fused connection units 78, Flex outlets 79

Wiring a shower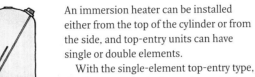

An electrically heated shower unit is plumbed into the mains water supply. The flow of water operates a switch to energize an element that heats the water on its way to the shower sprayhead. Because there's so little time to heat the flowing water, instantaneous showers use a heavy load – from 6 to 10.8kW. Consequently, an electrically heated shower unit has to have a separate radial circuit, which must be protected by a 30 milliamp RCD. In addition, for showers up to 10.3kW the radial circuit must be protected by a 45amp MCB or fuse, either in a spare fuseway at the consumer unit or in a separate single-way consumer unit; a 10.8kW shower needs a 50amp MCB. The circuit cable needs to be 10mm² two-core-and-earth.

The shower unit itself has its own on/off switch, but there must also be a separate isolating switch in the circuit. This must not be accessible to anyone using the shower, so you need to install a ceiling-mounted 45amp double-pole pull-switch (a 50amp switch is required for a 10.8kW shower). The switch has to be fitted with an indicator that tells you when the switch is 'on'.

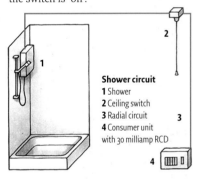

Shower circuit
1 Shower
2 Ceiling switch
3 Radial circuit
4 Consumer unit with 30 milliamp RCD

Connect the live and neutral conductors from the consumer unit to the switch's 'Mains' terminals, and those of the shower cable to the 'Load' terminals. Connect both earth wires, which have to be covered with green-and-yellow sleeving, to the single earth terminal on the switch.

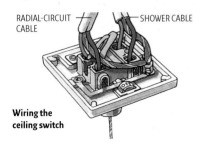

RADIAL-CIRCUIT CABLE   SHOWER CABLE

Wiring the ceiling switch

The shower unit itself must be wired according to the manufacturer's instructions. The unit and all metal pipes and fittings must be bonded to earth.

Immersion heaters

The water in a storage cylinder can be heated by an electric immersion heater, providing a central supply of hot water for the whole house. In many centrally heated homes the water is heated indirectly by the boiler, and an immersion heater is used as a backup for when the central heating has been switched off.

Types of immersion heater

1 Single element

2 Double element

3 Side-entry elements

An immersion heater can be installed either from the top of the cylinder or from the side, and top-entry units can have single or double elements.

With the single-element top-entry type, the element extends down almost to the bottom of the cylinder, so that all of the water is heated whenever the heater is switched on (**1**).

For economy, one of the elements in the double-element type is a short one for top-up heating, while the other is a full-length element that heats the entire contents of the cylinder (**2**). A double-element heater that has a single thermostat is called a twin-element heater; one with a thermostat for each element is known as a dual-element heater.

Side-entry elements are of identical length. One is positioned near to the bottom of the cylinder, and the other a little above halfway up (**3**).

Adjusting the water temperature
The thermostat that controls the maximum temperature of the water is set by adjusting a screw inside the cap covering the terminal box.

Adjusting the thermostat

Heating water on the night rate

If you have storage heaters, your electricity company supplies you with cheap-rate power for seven hours sometime between midnight and 8.00 a.m., the exact period being at the discretion of the company. This scheme is called Economy 7.

If you have a cylinder that is large enough to store hot water for a day's requirements, you can benefit by heating all your water during the Economy 7 hours. For the water to retain its heat all day, the cylinder must be insulated.

If your cylinder is already fitted with an immersion heater, you can use the existing wiring by fitting an Economy 7 programmer, a device that will switch your immersion heater on at night and heat up the whole cylinder. Then if you run out of hot water during the day, you can always adjust the programmer's controls to boost the temperature briefly. You can make even greater savings if you have two side-entry immersion heaters or a dual-element one. The programmer will switch on the longer element, or the

bottom one, at night; if the water needs heating during the day, then the upper or shorter element is used.

Economy 7 without a programmer
You can have a similar setup without a programmer if you wire two separate circuits for the elements. The upper element is wired to the daytime supply, while the lower one is wired to its own switchfuse unit and operated by the Economy 7 time switch during the hours of the night-time tariff only. A setting of 75°C (167°F) is recommended for the lower element, and 60°C (140°F) for the upper one. If your water is soft or your heater elements are sheathed in titanium or incoloy, you can raise the temperatures to 80°C (175°F) and 65°C (150°F) respectively without unduly reducing the life of the elements.

Leave the upper unit switched on permanently. It will only heat up if the thermostat detects a temperature of 60°C (140°F) or less, which should happen rarely if the cylinder is properly insulated.

SEE ALSO > Building Regulations on electrical wiring 70, Bonding to earth 72, Zones for bathrooms 73, Plumbing a shower 43

The circuit

The majority of immersion heaters are rated at 3kW; but although you can wire most 3kW appliances to a ring circuit, an immersion heater is regarded as using 3kW continuously, even though rarely switched on all the time. A continuous 3kW load would seriously reduce a ring circuit's capacity, so immersion heaters must have their own radial circuits.

The circuit needs to be run in 2.5mm² two-core-and-earth cable protected by a 15amp fuse or 16amp MCB. Each element must have a special double-pole isolating switch mounted near the cylinder; the switch should be marked 'water heater' and have a neon indicator. A 2.5mm² heat-resistant flexible cord runs from the switch to the immersion heater.

If the cylinder is situated in a bathroom, the switch must be outside zones 0 to 2. If this precludes a standard water-heater switch, fit a 20amp ceiling-mounted pull-switch with a mechanical on/off indicator. When installing any electrical appliance in a bathroom, the circuit must be protected by a 30 milliamp RCD.

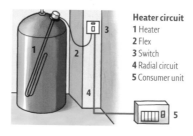

Heater circuit
1 Heater
2 Flex
3 Switch
4 Radial circuit
5 Consumer unit

20amp switch for an immersion heater

N If you undertake work marked with this symbol, you must inform the BCO before starting – see FIRST THINGS FIRST.

Wiring the switch and heater N

Feed the circuit cable into the switch mounting box fixed to the wall, and connect it to the 'Mains' terminals – brown to L, blue to N. Sheathe the earth wire in a green-and-yellow sleeve, then connect it to the common earth terminal (**1**). Prepare a heat-resistant flex for the switch. Connect the green-and-yellow earth wire to the common earth terminal and the other wires to the 'Load' terminals – brown to L, and blue to N. Tighten the flex clamp before screwing the switch faceplate in place.

The flex from the switch goes to the heater. Feed it through the hole in the cap, and then prepare the wires for connection.

Connect the brown flex wire to one of the terminals on the thermostat (the other one is already connected to the

wire running to the live terminal on the heating element). Connect the blue wire to the neutral terminal, and the green-and-yellow wire to the earth terminal (**2**). Replace the cap, which covers all the heater terminals and the thermostat.

Connecting to the consumer unit

Run the circuit cable from the cylinder cupboard to the fuse board. With the power switched off, connect the cable to an empty fuseway in the consumer unit. Although the consumer unit is switched off, the cable between the main switch and the meter will remain live – so take special care.

Having ruled out any obvious faults yourself, ask the BCO to carry out the necessary tests before you switch on and use the new circuit.

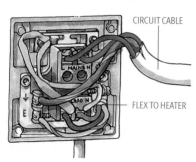

CIRCUIT CABLE

FLEX TO HEATER

1 Wiring the switch

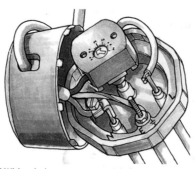

2 Wiring the heater

Replacing an element

With time, immersion heaters simply burn out and you are left without hot water. You can replace a heater yourself without notifying your BCO. First, check the circuit fuse or MCB; and if that isn't the source of the problem, isolate the heater circuit at the consumer unit and turn off the switch beside the cylinder. If your water is also heated by a boiler, switch that off, too.

Testing the immersion heater

Take the cap off the heater and, to test that the power has been switched off, touch the neutral terminal with one probe of a mains-voltage tester and the incoming live terminal on the thermostat with the other probe (**1**). Now do the same between incoming live terminal and the earth. If the power is off, make a note of the way the wires are connected to the heater, then disconnect the flex.

To check for a faulty thermostat, set it for maximum temperature and, using a multimeter set to its lowest resistance range, place one of the meter's probes on each of the thermostat terminals (**2**). If the meter shows no continuity, you only need to replace the thermostat – which saves having to drain the cylinder. If the thermostat seems to be functioning, place the probes on the heater terminals (**3**); and if there is no continuity, replace the heater.

Replacing the heater

To buy a replacement, estimate the diameter of the cylinder or the length of a top-entry heater. Heaters are usually supplied with the thermostat ready-fitted. You will also need to buy either an immersion-heater ring spanner or a special box spanner for turning an element surrounded by foam insulation.

Before you drain the cylinder, turn the heater very slightly with the spanner to free the threads (**4**). Now drain the cylinder, unscrew the heater and lift it out.

Fit the large washer supplied with the new heater, then screw the heater in place by hand until you feel the threads turning smoothly. The washer should prevent any leaks, but you can wrap PTFE tape around the heater threads as an extra precaution. Never smear it with sealant. Give a final turn with the spanner to tighten the heater, but don't apply too much force or you could distort the thin metal of the cylinder.

With the help of your notes, replace the wires on their terminals as they were on the old element (**5**); and then set the thermostat to the required temperature. Replace the cap, refill the cylinder and check for leaks; then restore the power.

1 Check the power is

2 Test the thermostat

3 Test the heater

4 Free the threads

5 Replace the wires

SEE ALSO > Draining the system 8, Building Regulations on electrical wiring 70, Zones for bathrooms 73, Testing circuits 74–5, Switching off the power 76

Index